English for Academic Research

More information about this series at http://www.springer.com/series/13913

Adrian Wallwork

English for Academic Research: Grammar, Usage and Style

Springer

Adrian Wallwork
Via Carducci 9
56127 Pisa, Italy
adrian.wallwork@gmail.com

ISBN 978-1-4614-1592-3 ISBN 978-1-4614-1593-0 (eBook)
DOI 10.1007/978-1-4614-1593-0
Springer New York Heidelberg Dordrecht London

Library of Congress Control Number: 2012948775

Printed on acid-free paper

Springer is part of Springer Science+Business Media (www.springer.com)

Introduction

Who is this book for?

This book is aimed at anyone who writes research papers, whose first language is not English, and who needs guidance regarding the grammar, usage, and style of academic English. It should also be useful for those who edit or proofread research papers.

It is suitable for those whose level of English is mid intermediate or higher.

How is the book organized?

The book is organized into 28 chapters on various aspects of English usage. This means that in the same chapter you will find all issues related to, for example, the use of tenses. However, some grammatical items are separated for convenience. For example, you will find the use of adverbs in three different sections: Chapter 13 deals with how adverbs are used as link words, Chapter 14 with typical differences between the various adverbs of time and place, and Chapter 17 with where adverbs are located within a phrase.

In each subsection, you will first find numbered guidelines. When there are four or more rules, the last few rules are generally the least important.

In the table below the guidelines, there are examples of sentences that implement (or fail to implement) the rules.

Good and bad examples are contained in the columns of the table entitled 'Yes' and 'No', respectively. The 'No' examples indicate typical mistakes taken from drafts of real papers. Most of the 'Yes' examples have been adapted from sentences written by native English speakers.

Sometimes in alternative to 'Yes' and 'No', one column is devoted to how one word or tense is used, and the other to how a related word or tense is used.

There is an index at the end of the book to help you find the particular grammar or style issue that you wish to review.

By consulting this book, will I improve my chances of getting my papers published?

Most definitely. This book is based on more than 25 years of experience of correcting the English of research papers. Guidelines and rules for avoiding around 2,000 typical mistakes are given. I have also read hundreds of referees' reports to understand what they pinpoint as being errors with regard to the English of a manuscript.

Does the book cover every aspect of English usage?

No, it only covers those problems that are generally found in research papers. For example, in this book the usage of tenses is explicitly related to how these are used in a paper, not to how they are used in the general spoken language. The only exception to this is the occasional example taken from 'general' English, where a 'general' example gives a clearer idea of usage than a 'scientific' example would do (this I have done above all in the section on modal verbs).

Aspects which are common to research papers and general English, but whose usage is very similar, are not covered in this book. For such issues, readers should refer to a general English usage guide such as Michael Swan's *Practical English Usage* (Oxford University Press).

By only focusing on those problems that typically arise in a research paper, I have been able to be more detailed in my explanations; for example, there is much more space dedicated to the use of articles (*a, the*, zero article) and the genitive than in other books. I have also been able to explain – I hope – the subtle differences between the present simple and past simple with regard to presenting results. I believe that no other grammar books have attempted to define these differences.

If, after reading a particular guideline in this book, you are still not completely sure how a word or grammatical construction is used, then try Google Scholar. This search engine trawls through thousands of manuscripts written by academics. It is thus a good way to check that you have correctly used, for example, an article (*a, an, the*), a preposition (e.g., *in, into, inside*) or collocation (i.e., a combination of words in a specific order). One good trick is to limit your search to those articles written by native speakers of English. To do this, type in 'Smith' in the 'author' field. Smith is the most common English surname and there are thousands of researchers with this surname. For more suggestions on how to use Google search engines, see Chapter 12 of the companion volume *English for Interacting on Campus*.

To what extent are the rules given in this book 100% applicable in all cases?

While doing my research for this book I analyzed papers written in a wide range of disciplines. What I discovered confirmed that each discipline (and indeed subdiscipline) tends to use English in very specific ways that are not consistent across disciplines.

An obvious example is the use of *we*. In some disciplines, *we* (and even *I*) are used freely; in other disciplines, they are explicitly banned. Less obvious examples are connected with the use of articles – where in one discipline the use of *the* and *a*/*an* in specific cases would be mandatory, in others it is not. Another example: the rules of punctuation, particularly regarding abbreviations and measurements, vary considerably from author to author, and journal to journal.

The exact rules of the genitive also seem to be impossible to define. At one stage of writing this book, I had written 16 pages on the use of the genitive, but in the end I decided that I was more likely to confuse rather than help my readers! So throughout I have tried to focus on the main areas that cause trouble rather than analyzing every conceivable case.

Being aware of this general lack of consistency in English use in academic writing is particularly important for those whose job it is to revise the English of manuscripts. Editors and proofreaders may find examples of constructions and word usage in the paper they are revising that go against the normal rules of English, but be unaware of the fact that this apparent misusage is perfectly acceptable in that particular discipline.

This is why I prefer to use the term 'guideline' rather than 'rule'. They are also very much *my* guidelines. Often they are based on my own intuitions derived from having read thousands of manuscripts, rather than any specific rules that I have been able to find in other grammar books or on the Internet. One big problem is that even native speakers within the exact same discipline do not always follow the same rules!

In this sense, this book should be seen as a draft of an ongoing project aimed at recording how the English language is used in academia. Please send your feedback to: adrian.wallwork@gmail.com

Other books in this series

English for Academic Research: A Guide for Teachers – tips for experienced EAP, ESP and Scientific English teachers, plus ideas on how to exploit the Writing and Presentation books in the *English for Academic series*.

English for Writing Research Papers – everything you need to know about how to write a paper that referees will recommend for publication.

English for Academic Correspondence – tips for responding to editors and referees, networking at conferences, understanding fast-talking native English speakers, using Google Translate, and much more. No other book like this exists on the market.

English for Interacting on Campus – tips for: socializing with fellow students, addressing professors, participating in lectures, improving listening skills and pronunciation, surviving in a foreign country.

English for Academic Research: Grammar/Vocabulary/Writing Exercises – these three books of exercises practice the rules and guidelines given in this manual. There are also writing exercises that can be combined with chapters from *English for Writing Research Papers*.

Cross-referencing between books

See http://www.springer.com/series/13913 to learn how this book can be used in combination with the other books in this series.

Recommended reading

I recommend the following books to learn more about non-language elements, e.g. how to produce bibliographies, cite the literature within the main text, and create tables and figures, plus more information on the use of measurements. Much of this information can also be found on Wikipedia.

A Manual for Writers of Research Papers, Theses, and Dissertations, Kate L. Turabian, the University of Chicago Press

MLA Handbook for Writers of Research Papers, Modern Language Association

Handbook of Writing for the Mathematical Sciences, Nicholas J. Higham, SIAM

Contents

The original of the book was revised in the front matter and in each chapter opening pages. An erratum can be found at DOI 10.1007/978-1-4614-1593-0_29.

1 Nouns: plurals, countable versus uncountable

1.1 regular plurals

1. To form the plural of most countable (1.6) nouns (including acronyms) simply add *s* or *es* to the end of the word.
2. In a *noun* + *of* + *noun* construction where the two nouns indicate a single entity, the first noun is made plural.
3. Adjectives are never made plural.
4. Nouns that act as adjectives are not made plural.
5. A noun which follows a number (or an implied number) is used in the singular form when acting as an adjective. Note the use of hyphens (25.6).
6. *-fold*, which is a suffix to indicate a specified number of parts or times, does not have a plural *-s*. Note the use of hyphens (25.6).

	YES	NO
1	We tested the engines of three car**s**, two taxi**s**, six train**s**, and four bus**es**.	
2	Several **points of view** have been put forward in the literature.	Several **point of views** have been put forward in the literature.
3	We also analysed three **other** papers on this topic.	We also analysed three **others** papers on this topic.
4	**Car** production is rising, but **car** sales are falling.	**Cars** production is rising, but **cars** sales are falling.
	= The production of cars is rising but the sales of cars are falling.	
5	I have a 24-**year**-old student helping me in the lab.	I have a 24-**years**-old student helping me in the lab.
	= The student is 24 **years** old.	
5	This work is part of a three-**phase** study into psychotic behavior amongst TEFL teachers.	This work is part of a three-**phases** study into psychotic behavior amongst TEFL teachers.
5	This would require a multi-**megabyte** memory.	This would require multi **megabytes** memory
6	The increase was **3-fold**.	The increase was **3 folds**.
	= There was a **3-fold** increase.	There was a **3 folds** increase.

A. Wallwork, *English for Academic Research: Grammar, Usage and Style*,
DOI 10.1007/978-1-4614-1593-0_1, © Springer Science+Business Media New York 2013

1.2 irregular plurals

1. Some nouns have irregular plurals: *child / children, man / men, woman / women, half / halves, knife / knives, life / lives, foot / feet, tooth / teeth.*
2. *Fish* and *sheep* are not made plural.
3. The plural of *mouse* (the animal) is *mice*, for the computer device the plural is *mouses*.
4. *Data* can be followed by the singular or plural – the plural form is more common in science. The singular form of data is *datum*, but *data* is more commonly used in both the singular and plural.
5. Datum / Data is an example of a Latin singular and plural. Other Latin and Greek words commonly used in scientific English are: *apex / apices, axis / axes, analysis / analyses, criterion / criteria, lemma / lemmata, optimum / optima, phenomenon / phenomena, vertex / vertices.*

	YES	NO
1	The patients consisted of three **children**, four adult **men**, and six adult **women**, all with persistent problems with their **teeth**.	The patients consisted of three **childs**, four adult **mans**, and six adult **womans**, all with persistent problems with their **tooths**.
2	This paper compares the relative brain powers of **fish** and **sheep**.	This paper compares the relative brain powers of **fishes** and **sheeps**.
3	All subjects were provided with PCs, monitors, headphones and **mouses**.	All subjects were provided with PCs, monitors, headphones and **mice**.
4	**This data is / These data are** inconsistent.	
5	This was true of the first analysis, but not of the other **analyses**.	This was true of the first analysis, but not of the other **analysises**.

1.3 nouns ending in -*s*

Some singular nouns finish in 's'. Such words behave in different ways:

1. *Economics, electronics, mathematics, physics, politics, statistics* – when these words describe a subject of study, they require a verb in its singular form (e.g. *is* not *are*).
2. If the words in Rule 1 are not used in the sense of a subject of study, they generally require the verb in the plural, but are also found with a verb in the singular. An exception is *electronics* which is found, indifferently, with a singular or plural verb.
3. *Means* can be the plural of *mean* (i.e. average). However, *means* is singular when the meaning is *way*, for example, *a means of transport.*
4. *News* is uncountable (1.8), also medical words such as *diabetes, mumps,* and *pus* are uncountable.
5. Nouns that end in *-is* form their plural with *-es* (e.g. *one analysis / thesis, two analyses / theses*).
6. *Species* is both singular and plural.

	YES	YES
1	**Economics is** one of the most popular subjects amongst students in our university.	
2	**Statistics is** a distinct mathematical science, rather than a branch.	It is not clear where **these statistics come** from.
2	In this case **the physics are** Eulerian invariant.	If **the physics is** the same in central and peripheral collisions, then Eq. 1 yields …
2	Competition is different in knowledge-based industries, because **the economics are** different.	Climate change is a subject of vital importance but one in which **the economics is** fairly young.
3	**This means** of transport **is** the fastest.	Prison is **another means** of controlling young offenders.
4	**This news is** not good.	
5	In my **thesis** I conducted an **analysis** of …	In their **theses** they conducted several **analyses** of …
6	Genome transplantation in bacteria: changing **one species** to another	**These species are** subdivided into serotypes.

1.4 nouns indicating a group of people

1. Some nouns that have a plural form are often used in the singular but with either a singular or a plural verb. Such nouns all relate to humans and include: *army* (*navy, air force*), *audience, board, cabinet* (*council, government, senate* etc.), *class* (as in group of students), *committee, company* (*firm, corporation* etc.), *crew, department, faculty, family, jury, majority, media, minority, public, staff, team*. The choice of singular or plural depends on whether the people who make up the group are acting as individuals (generally plural verb preferred) or as a collective unit (generally singular verb).

2. *People* requires a plural verb. *persons* is often used as a more formal version of *people*. *persons* is frequently found in medical and psychology research papers, or when talking about the capacity of a machine to hold a certain number of persons. In other cases *people* is often more appropriate particularly when it refers to people in general, rather than a subset.

3. *Police* is followed by a verb in the plural (e.g. *the police* do *not intervene*).

	YES	NO
1	The class **is** made up of 15 students.	The class **are** made up of 15 students.
1	The board of examiners **is / are** authorized to make decisions regarding …	The board of examiners **are** a statutory body established by the department.
2	Under pressure, **many people admit** that they believe in ghosts.	Under pressure, **much people admits** that they believe in ghosts.
2	Title: Prevention of heart disease in older **persons**	
	Title: A hypnotherapy treatment for **persons** prone to criminal activities	
2	Title: Job satisfaction – How do **people** feel about their jobs?	Title: Job satisfaction – How do **persons** feel about their jobs?
3	The police **are** often perceived as being racist.	The police **is** often perceived as being racist.

1.5 number-verb agreement

- Generally speaking the noun closest to the verb determines whether the verb is in a singular or plural form. Example: *The majority of* ***books have*** *now been digitized by Google.* In this example there are two nouns – *majority* and *books* – but *books* is closest to the verb (*have been digitized*).
- *A number of* requires a verb in its plural form; *the number of* requires a verb in its singular form.
- *A set of* or *a series of* requires a verb in its singular form.
- The verb before *more than one* is in its singular form.

	YES	NO
1	Around 40% of the **funds have** been deposited.	Around **40%** of the funds **has** been deposited.
1	The **majority** of **those** interviewed **were** African **Americans**.	The **majority** of those interviewed **was** African Americans.
1	Only a quarter of **these men are** still alive.	Only a **quarter** of all these men **is** still alive.
2	**A number of papers have** highlighted this major difference.	**A number of papers has** highlighted this major difference.
2	**The number** of papers being published on this topic **has increased.**	**The number** of papers being published on this topic **have increased.**
3	**A set** of three parameters **is** obtained.	**A set** of three parameters **are** obtained.
3	**A series** of four experiments **was** performed.	**A series** of four experiments **were** performed.
4	This happens when there **is more than one** possible answer.	This happens when there **are more than one** possible answer.

1.6 countable nouns: use with articles

A countable noun is something you can count: 30 books, many manuscripts, 100 apples, several PCs.

1. Before a singular countable noun you must put an article (*a / an* or *the*). For exceptions see 1.7.4.
2. If you are talking about something in general, then do not use *the* with plural nouns.
3. Scientific / technical acronyms (22) whose last letter stands for a countable noun behave like other countable nouns. They thus require an article when used in the singular, and an *-s* when used in the plural (22.3).
4. After *as* and *in*, a few singular countable nouns are used without any article.

	YES	NO
1	**A book** is still an excellent source of information.	**Book** is still an excellent source of information.
1	**The book** that I am reading is about …	**Book** that I am reading is about …
1	This acts as **an alternative**	This acts as **alternative**
1	When I was **a student**.	When I was **student**.
1	You cannot leave **the country** without **a passport**.	You cannot leave **country** without **passport**.
2	**Funds** are essential for research.	**The funds** are essential for research.
2	Throughout the world, **full professors** tend to earn more than **researchers**.	Throughout the world, **the full professors** tend to earn more than **the researchers**.
3	Access requires **a PIN** (personal identification number).	Access requires **PIN** (personal identification number).
3	The number of purchases of **CDs** is only 1% of what is was 25 years ago.	The number of purchases of **CD** is only 1% of what is was 25 years ago.
4	We used a 5-kR resistor placed **in series**.	We used a 5-kR resistor placed **in a series**.
4	All non dummy variables are **in log form**.	All non dummy variables are **in a log form**.
4	We used X **as input**, and Y **as output**.	We used X as an input, and Y **as an output**.

1.7 singular countable nouns: use with and without *a / an* in scientific English

1. Some singular countable nouns can be used with or without an article with no difference. There are no clear rules for this, and usage seems to vary from discipline to discipline, and from author to author.
2. If the noun is followed by *of* (i.e. to add further details), then this noun is preceded by *a / an*.
3. Some singular countable nouns are used without an article when they are used in an extremely generic way.
4. When preceded by *by*, means of transport are used without *a / an*; certain time expressions do not require *a / an* when used with prepositions.

	WITH A / AN	WITHOUT A / AN
1	It is stored in **a compact form**	It is stored in **compact form**.
1	As these parameters are fixed, a grammar is determined, what we call **a "core grammar"**	We call this kind of abstraction "**aggregation.**"
1	These were obtained by using 3-chloro-1-propanol **as the internal standard**.	These fats were used **as internal standard**.
1	**An analysis** of the data showed that …	**Analysis** of the data showed that …
1	… with **a probability** of 0.25	… with **probability** 0.25.
1	The software is used under **a license** from IBM.	The software is used under **license** from IBM.
2	This analysis indicated that the number of strata could be reduced considerably **without a loss** in the precision of the values found.	This analysis indicated that the number of strata could be reduced considerably **without loss** of precision and **without loss** of generality.
2	The guinea-pigs were housed singly or in pairs at **a room temperature** of 20–22°C.	The samples were stored at **room temperature.**
2	This was followed by etching in **an aqueous solution** of phosphoric acid and chromic acid.	We examined the reaction between methylchloride and chloride ion in the gas phase and in **aqueous solution** using techniques based on …
3	Their new perfume depicts **a strawberry** on the label.	Their new perfume smells of **strawberry**.
4	They rented **a car** to travel through India.	They traveled through India **by car**. They drove **by night**. They discovered that it often rains in India **in [the] summer**.

1.8 uncountable nouns: general rules

An uncountable noun is seen as a mass rather than as several clearly identifiable parts, for example chemicals, gases, metals, and materials. There are hundreds of uncountable nouns, some examples frequently used in research are:

access, accommodation, advertising, advice, agriculture* (and other subjects of study), *capital, cancer* (and other diseases and illnesses), *consent, electricity* (and other intangibles), *English* (and other languages), *equipment*, evidence*, expertise, feedback, functionality, furniture*, gold** (and other metals), *hardware, health, industry, inflation, information*, intelligence, luck, knowhow, luggage*, machinery*, money, news, oxygen* (and other gases), *personnel, poverty, progress, research, safety, security, software, staff, storage, traffic, training, transport, waste, wealth, welfare, wildlife.*

The uncountable nouns listed above with an asterisk (*) can be used with *a piece of*. This means that they can be used with *a / an, one* and be made plural. Examples: *a piece of advice, two pieces of equipment, one piece of information.*

Uncountable nouns cannot be:

1. Made plural, i.e. you cannot put an 's' at the end of the word; this means that they are not used with plural verbs (e.g. *are, have*).
2. Preceded with words such as: *a, an, one, many, few, several, these* (i.e. words that in some way indicate that a distinct number of items is involved).

	YES	NO
1	**This information is** confidential.	**These informations are** confidential.
1	**Feedback** from users on usage of the software **has** shown that …	**Feedbacks** from users on usage of the software **have** shown that …
1	The **news is** good – our manuscript has been accepted.	The **news are** good – our manuscript has been accepted.
2	We need **several new pieces new equipment** and **[some] new software**.	We need **several new equipments** and **a new software.**
2	Our institute only has **a little money** available for funding.	Our institute only has **few money** available for funding.
2	We have not done **much research** in this area.	We have not done **many researches** in this area.
2	**Written consent** was obtained from all patients.	**A written consent** was obtained from all patients.
2	She has **expertise** in this field.	She has **an expertise** in this field.

1.9 uncountable nouns: using a different word or form

1. To express the plural of certain uncountable words, sometimes you need to choose another word.
2. On other occasions you may need to place the uncountable noun in an adjectival position before another noun.

	YES	NO
1	She is **an expert** in many areas.	She has **expertises** in many areas.
1	The **features** of this application **are** outstanding.	The **functionalities** of this application are outstanding.
	The **functionality** of this application **is** outstanding.	Note: Although theoretically uncountable, *functionalities* is gaining acceptance
1	They have **a new advertisement** on TV.	They have **a new advertising** on TV.
1	I have done **several jobs** both in industry and research.	I have done **several works** both in industry and research.
1	They work in research and also for **a manufacturing company**.	They work in research and also for **an industry**.
1/2	We need **a program** / **an app.**	We need **a software.**
	We need **a software application**.	
2	We have **a training course** tomorrow.	We have **a training** tomorrow.

1.10 uncountable nouns: more details

1. Some nouns are both countable and uncountable, but with a difference in meaning.
2. Some nouns are used in both their singular and plural forms, with no difference in meaning.
3. Some uncountable nouns can be used in a countable way when preceded by an adjective.

	UNCOUNTABLE	COUNTABLE / PLURAL FORM
1	**Paper** and **coffee** are becoming expensive commodities.	She has **a coffee** (i.e. a cup of coffee) and reads **a paper** (i.e. a newspaper) every day.
		She has just finished **another paper** (i.e. a manuscript)
1	The role of traditional medicine is being undermined by alternative **medicine**.	The occurrence and fate of **medicines** in the environment – i.e. how they are absorbed into the water and soil systems – has rarely been investigated.
1	The explosion caused considerable **damage** to the machine.	The company has been awarded **damages** (i.e. compensation) as a result of the lawsuit.
1	Dealing with **waste** is a major problem in the West.	The conference was **a waste of time**.
1	This **work** (i.e. this research, manuscript) is worth publishing.	The field of the cultural heritage investigates ways of preserving **works** of art.
2	This **data is** fascinating.	These **data are** fascinating.
2	Teenagers often exhibit **behavior** that is annoying for adults.	Some autistic children exhibit **behaviors** that are potentially …
2	Several devices were tested and their **performance** was evaluated.	Several devices were tested and their **performances** were evaluated.
3	This does not imply prior **knowledge** of …	She has **a good knowledge** of English.

2 Genitive: the possessive form of nouns

The rules for when to use ‘s to indicate possession are not clear and are often contradictory. Even native speakers are inconsistent, though most intuitively know what is and is not correct.
The rules of general English are that you should only use the genitive with:

- people, companies, insitutes etc (e.g. *Smith's book, Apple's profits, IMT's staff*)
- animals (e.g. *the dog's bone*)
- in certain time expressions (e.g. *in three years' time*)

If you are not sure whether to use the genitive first see if you can find similar examples using Google Scholar. If you are still not sure then use the following formula: noun + *of* + noun (e.g. *the assets of the company* rather than *the company's assets*).

In any case, if you misuse the genitive it will rarely constitute a serious mistake. This section details when the above rules are and are not respected in research manuscripts.

A. Wallwork, *English for Academic Research: Grammar, Usage and Style*,
DOI 10.1007/978-1-4614-1593-0_2, © Springer Science+Business Media New York 2013

2.1 position of the *'s* with authors and referees

1. The *'s* is placed immediately after the last letter of the author (or name, country, etc.). Note: do not use *the* before the name of the author.
2. Even if the last letter of the author's name is an *s*, then still put an *'s*. Exceptions: non-English surnames that end in a silent *-s* (e.g. *Camus' first novel, Descartes' meditations*).
3. When a paper has been written jointly by two authors, only put an *'s* after the last name or after et al. A similar rule applies to compound nouns (e.g. *his mother-in-law's house*).
4. If two papers were written by two authors separately, then the *'s* must be used for both authors.
5. If the noun is in the plural (e.g. referees, those authors, editors), then put just an apostrophe (i.e. no *s*) after the plural *-s*.
6. When a referee is referred to by a number, put the *'s* after the number.

	YES	NO
1	**Simpson's** paper is an excellent introduction to the topic.	**The Simpson's / Simpson** paper is an excellent introduction to the topic.
1	We have answered the **referee's** questions	We have answered the **referee** questions.
	there is just one referee involved	
1	I have just received the **editor's** decision along with the **committee's** report.	I have just received the **editor** decision along with the **committee** report.
2	**Jones's** seminal paper.	**Jones'** seminal paper.
3	**Smith and Simpson's** paper.	**Smith's and Simpson's** paper.
3	**Smith et al's** paper.	**Smith's** et al paper.
4	**Smith's paper and Li's paper** take two very different positions.	**Smith and Li's paper** take two very different positions.
5	It is each applicant's responsibility to ensure that the **three Referees'** Reports are submitted by …	It is each applicants' responsibility to ensure that the **three Referee's** Reports are submitted by …
6	We have answered the three referees' questions, and specifically, we have added a new section as per **Referee 1's** request.	We have answered the three referees' questions, and specifically, we have added a new section as per **Referee's 1** request.

2.2 theories, instruments etc.

1. Do not use *the + name of person + 's.*
2. *The + name of person + noun:* this construction can be used instead of the genitive, with no change in meaning. This construction is very formal and is only used with famous scientists. This means that you cannot write ~~*the Adrian Wallwork theory of writing*~~, because Adrian Wallwork (the author of this book) is not sufficiently famous!
3. *Name of person + 's + noun*: the focus is usually (but see Rule 5) more on the scientist. We are talking about their original concept, their life etc.
4. *Name of person (used adjectivally) + noun*: when the focus is primarily on the use that the author of the paper has made of the scientist's method, rather than the focus being on the scientist himself / herself.

	GENITIVE	NO GENITIVE
1	**Adrian Wallwork's** manual on writing. ~~The **Adrian Wallwork's** manual on writing.~~	
2	As predicted by **Newton's theory** of gravity, Mercury's orbit is elliptical.	**The Newton Theory** of Gravity states that …
2	The premise of **Darwin's theory** of evolution is that …	This work was inspired by **the Darwin Theory** of Evolution.
3,4	**Fourier's analysis** of linear inequality systems highlights that **he** placed more importance on …	We used **Fourier analysis** to evaluate the …
3,4	**Turing's machine** was designed to be an idealized model of a human computer.	We may think of **a Turing machine** as a …
3,4	**George Boole's** father was a tradesman who gave his son his first lessons in logic and mathematics.	**Boolean algebra** is a logical calculus of …

2.2 theories, instruments etc. (cont.)

5. In some cases the genitive is used even when the focus is on how a scientist's theory or test was used by the author, rather than focusing on the scientist. Note *a + name of person + noun*: when making reference to pieces of equipment etc.
6. When a law, theory etc. was the invention of more than one scientist, then the *'s* only follows the name of the last scientist. Rule 2 can also be applied in such cases.
7. In some cases where two scientists are involved, the construction given in Rule 2 is preferred.

	GENITIVE	NO GENITIVE
5	One-way ANOVA with **Tukey's post hoc test** for individual treatment differences was used for statistical analysis.	**A Tukey post hoc test** was used to compare the four groups.
6	**Beer-Lambert's law** has often been used to model canopy transmittance.	**The Beer-Lambert law** has often been used to model canopy transmittance.
7		In this paper the **Kolmogorov-Smirnov statistical test** for the analysis of histograms is presented.

2.3 companies and politicians

Rules 2 and 3 in 2.2 also apply to companies and politicians.

	GENITIVE	NO GENITIVE
2,3	**Nike's** decision to raise the prices of their shoes is in direct contrast to **Camper's** decision to lower their prices.	The survey found that typical consumers had, over the 12-month period, bought at least two **Nike** products and one **Apple** *i*-phone or *i*-pad.
	Nike is seen here as a group of managers within a company.	Nike and Apple are used like adjectives to describe a product, the two companies are not being seen in terms of their managers.
2,3	**Obama's** administration was initially much more popular than Bush's or Clinton's.	The **Obama** tried to block Alabama's new administration immigration laws.
	Focus on the president contrasted with other presidents	Focus on all the people who worked for Obama seen as a whole

2.4 universities, departments, institutes etc.

1. High positions of people associated with universities etc. tend to be written without using the genitive.
2. Use *the + university + of + town* in formal situations (e.g. in prospectuses, on websites, in articles, in official documents).
3. Use *town + university* when we see things from the student's point of view. This construction is less formal, but in any case can always be replaced by the construction given in Rule 2.

	YES	NOT COMMON (1,2), WRONG (3)
1	The **Chancellor of the University of Cambridge** is meeting the **Rector of the University of Coimbra.**	The **University of Cambridge's chancellor** is meeting the **University of Coimbra's Rector.**
2	**The University of Bologna** is the oldest university in the world.	**Bologna University** is the oldest university in the world.
3	I studied at **Bologna University / the University of Bologna**.	I studied at **Bologna's University**.

2.5 animals

1. Use *'s* when referring to the parts of the body of a living animal.
2. Use *'s* when referring to the products of living animals.
3. Do not use *'s* for dead animal body parts or products.

	YES	NO
1	The temporal lobes of the **monkey's brain**.	The temporal lobes of the **monkey brain**.
2	We used **ewe's milk** rather than **cow's milk**.	We used **ewe milk** rather than **cow milk**.
2	**Lamb's wool** is ideal for this kind of outdoor clothing.	**Lamb wool** is ideal for this kind of outdoor clothing.
3	Collagen can be obtained from **calf skin** or **rat skin**.	Collagen can be obtained from **calf's skin** or **rat's skin**.
3	In some parts of the world they eat **monkey brain**.	In some parts of the world they eat **monkey's brain**.

2.6 genitive with inanimate objects

The genitive is not generally used with non-human subjects, apart from those categories mentioned in the previous subsections (companies, countries, towns, planets). However in some cases – for which there are no rules – the genitive is used with inanimate things. Its usage varies from discipline to discipline, and may break the usual rules of English grammar. In most cases a *the + noun + of + the + noun* construction can also be used. Thus if you are not sure, use the *of* construction. See also 16.10 and 16.11

YES (NEARLY ALWAYS CORRECT)	YES (BUT ONLY IN SOME CASES)
The role of the **brain** is crucial.	The **brain's** role is crucial.
The tasks of the **network** is to converge to a particular output.	The **network's** task is to converge to a particular output.
An understanding of **the effects of malaria** on the region's inhabitants is vital.	An understanding of **malaria's** effects on the region's inhabitants is vital.
The radius of the **circle**.	The **circle's** radius.
The approximate time of the arrival of the **plane** was calculated.	The approximate time of the **plane's** arrival was calculated.
The occupants of the **flat** were all arrested.	The **flat's** occupants were all arrested.

2.7 periods of time

1. The genitive is used when a time period is used adjectivally.
2. The genitive is not used when time periods are preceded by *a / the*. Note that the first noun in the noun + noun construction is in the singular form. This is because the first noun functions as an adjective to describe the second noun.

	YES	NO
1	I'm taking three week**s'** vacation next month.	I'm taking three **weeks** vacation next month.
	= three week**s** of vacation	
2	He's on a 3-wee**k** vacation.	He's on a three week**s'** vacation.
	He's on a three-wee**k** vacation.	He's on a three week**s** vacation.

3 Indefinite article: *a* / *an*

3.1 *a* versus *an*: basic rules

a is used before:

1. All consonants (but see Rule 8 below).
2. *U* when the sound is like *you* (e.g. *university, unique*).
3. *Eu* (but not in acronyms).
4. *One*.
5. *H*, except for the words listed in Rule 8 below.

an is used before:

6. *A, e* (but not *eu*) i, and o.
7. *U* when the sound is like the *u* in *understanding, unpredictable*.
8. *Hour, honor, heir, honest* and their derivatives, and *herb* / *herbicide* (US English). *an* is not used before other words that begin with H, unless the H appears in an acronym. Note: both *a* and *an* are commonly used before *historical*.

	A	AN
1,6	**a** Sony laptop, **a** Vodafone application	**an** Apple laptop, **an** Orange telephone
2,7	**a** universal law	**an** undisputed argument
3	**a** European project	**an** EU project
4	**a** one-off payment, **a** one-day trial	
5,8	**a** hierarchy, **a** Hewlett Packard computer	**an** hour, **an** HP computer

A. Wallwork, *English for Academic Research: Grammar, Usage and Style*,
DOI 10.1007/978-1-4614-1593-0_3, © Springer Science+Business Media New York 2013

3.2 *a* versus *an*: use with acronyms, digits, and symbols

1. Use *a* before the following letters in acronyms: B, C, D, G, J, K, P, Q, T, U, V, W, Y, Z.
2. Use *an* before the following letters in acronyms: A, E, F, H, I, L, M, N, O, R, S, X.
3. Sometimes acronyms are read as words (e.g. NATO, URL, PIN, UNICEF) rather than letter by letter (e.g. EU, UN, US). If they are read as words then the normal rules for *a* / *an* apply. If they are read as letters, then rules 1 and 2 apply.
4. When deciding between *a* or *an* before a number written in figures (e.g. *a 100 kilowatt battery*) say the word out loud in your head and follow the normal rules (e.g. *a one hundred kilowatt battery* follows Rule 4 in the previous subsection, *an eight kilowatt battery* follows Rule 7).
5. Before symbols and Greek letters decide whether the word that the symbol or letter represents would be used with *a* or *an*, following the rule of the previous subsection.

	A	AN
1,2	a US soldier, a VIP lounge, a YMCA hostel	an IBM machine, an MTV program, an SOS signal
3	a USB, a NATO officer	an url, an NLP course
4	a 1 GB disc, a 10 GB disc, a 12 GB disc	an 8 GB disc, an 11 GB disc, an 18 GB disc
5	a # (a hash)	an Σ (an epsilon)
	a % (a percentage)	an * (an asterisk)

3.3 *a / an* versus *one*

one is a number (*one, two, three*). Use *one* instead of *a / an*:

1. When it is important to specify the number.
2. Before *another.*
3. Before *way* when not preceded by an adjective.
4. In expressions of this type: *one day next week.*

	ONE	A / AN
1	We need **one** manual, not two manuals.	We need **a** manual, not just any type of document.
1	Unfortunately, there is only **one** solution in such cases – surgical intervention.	In this paper we present **an** innovative solution to the three-bus problem.
1		This parameter has **a unique** value.
1	If you make even **one** mistake with Prof Syko, she will fail you.	If you make **a** mistake with Prof Normo, it's not a problem – he's really relaxed.
1	We conducted **one** experiment in which students had to memorize 100 words in English, and another in which they had to remember 200 words.	We conducted **an** experiment in which students had to memorize 100 words in English. This was the only experiment we conducted and it proved that …
2	We went from **one** town to **another**.	The conference is in **a** town near Istanbul.
3	**One** way to do this is to …	**A novel** way to do this is …
4	We could have the meeting **one day** next month.	**A good day** to meet would be next Tuesday.

3.4 *a* / *an* versus *the*: generic versus specific

1. Use *a* / *an* first time you mention something.
2. Use *the* on subsequent occasions (i.e. when the reader / listener already knows what you are talking about).
3. Use *a* / *an* to refer to something generic, *the* to something specific or something which the reader will already be familiar with.

	A / AN	THE
1,2	The only thing you can take into the examination tomorrow is **a dictionary**.	The only thing you can take into the examination is a dictionary. **The dictionary** you choose can either be mono- or bi-lingual.
1,2	This paper presents **a new system** for modeling 4D maps.	This paper presents a new system for modeling 4D maps. **The system** is based on …
1,2	I don't have **a computer** at home.	I have a computer at home and at work. **The computer** that I have in my office is a Mac and the one at home is an HP.
1,2	ABSTRACT In this work, we make **an attempt** to test the efficiency of …	RESULTS **In this work, the attempt to** assess the relative efficiency of the tested methods was carried out on two levels.
3	**A comparison** of our data with those in the literature indicates that …	**The comparison** given in Sect. 2.1 highlights that …
3	We are now in **a position** to apply Theorem 13.	The diagram indicates **the position** of each piece of equipment.
3	Contrary to what is currently thought, there is **a growing demand** for experts in this field.	We need to satisfy **the growing demand** for experts in this field, which looks set to increase even further.
3	This is **a first step** towards combatting terrorism in that area. We cannot be sure of the outcome …	This is **the first step** towards combatting terrorism in that area. The second step is to …

3.5 *a / an* versus *the*: definitions and statements

1. Use *a / an* when talking about one example of a category (i.e. a division of people or things with similar characteristics). In such cases *a* means *any* (6.2, 6.3).
2. Use *the* to generalize about the entire set of components in a class. In such cases *the* means *all the.*
3. Use *a / an* in definitions.
4. Use *the* to make general statements about some entity.

	A / AN	THE
1, 2	**A camel** (= *any camel*) can go for days or even months without water because, unlike other animals, camels retain urea and do not start sweating until their body temperatures.	**The panda** (= *all the pandas in the world*) is in danger of becoming extinct.
3, 4	**A computer** is an electronic device for storing and processing data.	**The computer** has changed the way we live.

3.6 *a / an, the*, possessive pronoun: parts of the body

1. In definitions use *a / an* before external organs, and *the* before internal organs. *His / her / their* are more informal.
2. Use *a / an* for generic statements, *the* for specific cases. Only *his / her* when the body part belongs specifically to the male or female, respectively.
3. *A / An* is used when the person / animal has more than one of a particular body part, *the* is used when the part of the body is a unique item.
4. If the person or animal has many of the same body parts and you are referring to an individual item of such body parts use *a / an*. If you are referring to all of them use *the.*
5. *The* is used when someone has something inflicted on him / her, or when the body part is being focused on rather than the fact that this body part belongs to someone.

	A / AN	THE	HIS, HER, THEIR
1	**A beard** is the growth of hair on the face of an adult male.	**The heart** is the most important muscle of the human body.	**Your heart** is about the same size as your fist and weighs a little less than two baseballs.
2	The patient had camouflaged his abnormal neck appearance with **a beard**.	The average length of the long guard hairs of the goat near the front of **the beard** was measured.	Employees cannot be fired in cases where the employee refuses to shave **his beard**.
3	The patient, a male aged 24, had burned **an arm**.	The patient complained of discomfort in **the back**.	The patient complained of discomfort in **his back**. He had also burned **his left arm**.
4	When hexanol is placed on the antennae of an insect, the insect cleans itself. When it is held close to **an antenna**, the insect normally turns away.	Dust that might entangle **the antennae** of the parasites was removed with a small brush.	The **male** mounted the **female** and aligned himself along the axis of **her body**, and tried to place **his antennae** between those of the female.
5	We managed to relieve a patient of a pain in **a leg** that had been amputated several years before.	The bullet hit him in **the arm**. He was hit in **the arm**.	In the second year of **her** illness, the patient developed stiffness in **her arm**.

4 Definite article: *the*

4.1 definite article (*the*): main usage

The principle use of *the* is to refer to something specific (i.e. particular cases rather than all cases). However, the distinction between general and specific is not always straightforward, as illustrated by these two examples:

a) **Male professors** of physics from China who also work in the field of mathematics and how have studied in the USA, tend to … b) **The male professors** of physics who also work in the field of mathematics that **Anna met** at the conference are ...

Sentence (a) seems very specific - but it isn't. For something to be specific we have to be able answer the question "which one/s?". In (a) we don't know which specific professors. In the second sentence (b) the fact that Anna met them means that we are not referring to all such professors in the world, but a very specific subset of them, i.e. the ones that Anna met at the conference.

The examples below show typical cases where the definite article must be used in English, but where it may not be used in your language.

YES	NO
The aim of this document is to prove …	**Aim** of this document is …
Our aim	
The computers that are used in our department are all Hewlett Packard, and **the software** that we use is all proprietary software.	**Computers** used in our department are all Hewlett Packard, and **software** that we use is all proprietary software.
Our computers and software	
The government have increased taxes.	**Government** have increased taxes.
The government of our country	
As reviewed **in the literature** …	As reviewed **in literature** …
The literature in our field	
All the samples were cleaned in **the** laboratory.	All the samples were cleaned **in laboratory**.
The lab in our institute	
The results of the present study show …	**Results** of the present study show …
Our results	

A. Wallwork, *English for Academic Research: Grammar, Usage and Style*,
DOI 10.1007/978-1-4614-1593-0_4, © Springer Science+Business Media New York 2013

4.2 specific versus general: examples

The term 'specific' with reference to the definite article means that the noun is qualified in some way. Typical qualifications are:

1. Another noun: a *noun1* + *of* + *noun2* construction indicates that probably *noun1* is being specified by *noun2*. In such cases, *noun1* should be preceded by *the*.
2. A noun + (*that*) + (subject) + verb.
3. A superlative (19), e.g. *the best, the simplest.*
4. An adjective such as *first, second* (etc.), *main, principal, only, initial.*
5. Adjectives (even a whole sequence) don't necessarily make their noun specific.

	SPECIFIC	GENERAL
1	**The life of a peasant** in the Middle Ages was hard.	**Life** in the Middle Ages was hard.
1	**The history of English** is fascinating.	**History** was my favorite subject at school.
2	**The problems that we've been having** with our English pronunciation are very serious.	**Problems** when learning English are very common.
2	**The wheat** used in some types of food is derived from …	Studies were carried out on **wheat**.
2	**The hydrochloric acid** employed in our studies was purchased from …	**Hydrochloric acid** is twelve times more active than sulfuric acid.
3	This is **the worst** paper in the collection	**Poorly written manuscripts** are very common.
4	**The main differences** are: X, Y and Z.	**Differences** in opinions on this subject are very common.
5,2	**The red wine that we had** last night.	I prefer **dark red wine from Chianti** to sparkling white wine from Asti.
5,2	**The** intelligent female Ph.D. students from non-European countries who have studied English **that have attended** my course tend to get better results than …	Intelligent female Ph.D. students from non-European countries who have studied English tend to get better results than …

4.3 other uses of the definite article

Use *the*

1. With certain expressions: *the Internet, the weather, the sun, the environment, the dark.*
2. To indicate a class of objects in an abstract sense. Note: in a definition use *a / an* (e.g. *A computer is a machine that performs calculations*).
3. With *last* and *next* in time expressions to indicate a specific week, month, year etc. rather than the current week etc.

	YES	NO
1	We found your address on **the Internet**.	We found your address on **Internet**.
1	Samples were stored in **the dark** at room temperature.	Samples were stored **in dark** at room temperature.
2	**The computer** and **the telephone** have changed the way we live.	**Computer** and **telephone** have changed the way we live.
3	The conference has been organized for **the last week** in May.	The conference has been organized for **last week** in May.
3	We will be sending you our manuscript **next** week.	We will be sending you our manuscript **the next** week.

5 Zero article: no article

5.1 zero article versus definite article (*the*): main usage

The term 'zero article' refers to cases where no article is required. Use the zero article if you are talking about something in general and the noun is:

1. In the plural, e.g. *computers, books.*
2. Uncountable (1.8), e.g. *hardware, information.*
3. Abstract – either singular countable (1.6) or uncountable e.g. *life, success, performance.*

Note that:

4. Some words change meaning if they are used with or without *the.*
5. Titles to papers occasionally omit the article of the first noun. Both forms (i.e. with and without *the*) are common.
6. Captions to figures often omit the definite article.

	ZERO ARTICLE	THE
1	Oracle do not sell **computers**.	The **computers** that we have at our institute are …
2	Oracle sell **software**.	The most commonly used **software** is
2	**Research** is essential if progress is to be made.	**The research** that we have conducted so far proves that …
3	There was a significant effect of the road conditions on **speed**.	**The speed** of the car was optimal.
4	I love **nature**.	**The nature** of this problem is not clear.
4	The probe has been launched into **space**.	**The space** between A and B must be wide enough to accommodate C.
5	**Development** and validation of a test to measure competence in English	**The development** and validation of a group testing of logical thinking
6	Figure 1. **Average rainfall** 2010–2020.	We predicted **the average rainfall** for 2020.

A. Wallwork, *English for Academic Research: Grammar, Usage and Style*,
DOI 10.1007/978-1-4614-1593-0_5, © Springer Science+Business Media New York 2013

5.2 other uses of the zero article

1. In expressions containing *from … to*, e.g. *from top to bottom, from coast to coast.*

2. With names of public buildings and places when used to refer to their primary purpose (*he is a Ph.D. student, he studies at university*). These include: *school, university* (but not *department* or *institute*), *college, work* (but not *office*), *home, church, hospital, prison.*

3. Before the names of people, unless the name is being used adjectivally (2.2).

	ZERO ARTICLE	THE
1	Figure 5: From **left to right**, the Dean, the Dean's husband, and Prof. Donald Duck.	In GB they drive **on the left**, in the rest of Europe **on the right**.
2	Before going **to school** I was educated **at home**. I then **left school** at 18 and then went **to university**.	The editors also wish to record their thanks to **the School** of Sociology and Social Policy at **the University** of Leeds for its continuing support.
3	**Davidson's article** is important for several reasons.	This paper deals with **the Davidson method** which computes a few of the extreme eigenvalues of a symmetric matrix and corresponding eigenvectors.

5.3 nationalities, countries, languages

1. When talking in general, *the* must be used with 'uncountable' nationalities that end in – *h* (e.g. *English, French*) and *–ese* (e.g. *Chinese, Portuguese*). Most other nationalities (*Italians, Swedes* etc.) are countable and can be used with or without *the.*
2. If a nationality that ends in – *h* or *–ese* is found with another nationality, then for the sake of consistency all the nationalities are preceded by *the.*
3. Rule 1 above does not apply if these words are being used as adjectives rather than nouns, e.g. before *people, men, women.*
4. Continents and countries do not require the article: *Europe, Asia, Italy, France, Russia.* Exceptions: *the UK, the USA, the Ukraine, the United Arab Emirates, the ex-USSR, the Arctic, the Antarctic.*
5. Do not use *the* with languages when these languages are being talked about in general.

	ZERO ARTICLE	THE
1	**Italians** do it better than **Americans**.	**The English** are not as tall as **the Portuguese**.
2		**The English** are not as tall as **the Portuguese** or the **Italians**.
3,1	**Chinese people** are not as tall as **Japanese people**.	**The Chinese** are famous for their culture.
4	We have offices in **France, Spain** and **Italy**.	We have offices in **the UK** and **the USA, France, Spain** and **Italy**.
5	**The English** of this paper needs to be revised.	**English** is not an easy language to learn.

5.4 zero article and *the*: contradictory usage in scientific English

The second column in the table below lists some occasions where the normal rules of the use of articles in English have apparently been broken but are nevertheless frequently found in research papers written by native speakers.

NORMAL ACCEPTED USAGE	ALSO POSSIBLE IN SCIENCE
After **the incubation**, all complexes were analyzed on 0.8% agarose gels and electrophoresed in TBE.	After **incubation**, the number of bacteria was determined by a direct count.
The inhibition of this enzyme is thought to be responsible for the cytotoxicity of …	**Inhibition** of this enzyme by analogous chemical compounds has been found to decrease the proliferation of P. falciparum.
At present, **the annotation** of the proteins of *A. gambiae* is preliminary.	**Annotation** of the proteins of these new genomes can be transferred to closely related genomes.
Title: The effects of salinity on dry matter partitioning and fruit growth in **tomatoes** grown in nutrient film culture.	Title: Fruit Yield and Quality in **Tomato**
Title: Occurrence of flavonols in **tomatoes** and tomato-based products	Title: Identification of two genes required in **tomato**
Those compounds which have been most effective on wheat have invariably been proportionately active on **the tomato**.	In this study, we describe a recessive mutant of **tomato**.
Lycopene, found primarily in **tomatoes**, is a member of the carotenoid family.	

5.5 zero article versus *a* / *an*

1. *A* / *An* must be used before a singular countable noun (1.6), the zero article before an uncountable noun (1.8).
2. *A* / *An* must be used before names of instruments, pieces of equipment etc.
3. With reference to an academic position, *a* / *an* refers to a job that is held by several people. The zero article is used when stating a specific job position that is only held by one person.

	A / AN	ZERO ARTICLE
1	When I was **a student**, I was a member of the students' union.	The referees gave us **feedback** on our manuscript.
1	You cannot travel there without **a passport** or without **a visa**.	You cannot travel there without providing **information** about the reason for going.
2	**A** Thermoquest Trace GC gas chromatograph with a PTV injector and coupled with **an** ion trap mass spectrometer PolarisQ was used.	We used **equipment** located in our laboratory.
3	He is **an assistant professor** at the University of Seoul.	He is **Assistant Professor** of Pediatrics at the University of Seoul.
3	She is **a professor**, not **a senior researcher**.	She is **Professor of Education** at the University of Atago.

5.6 zero article and *a* / *an*: contradictory usage in scientific English

The second column in the table below lists some occasions where the normal rules of the use of articles in English have apparently been broken but such occurrences are nevertheless frequently found in research papers written by native speakers.

NORMAL ACCEPTED USAGE	ALSO POSSIBLE IN SCIENCE
An analysis of the data showed that …	**Analysis** of the data showed that …
A further analysis of the data showed that …	**Further analysis** of the data showed that …
A statistical analysis of the data showed that …	**Statistical analysis** of the data showed that …
We investigate natural products **of an animal origin**.	They include strains **of animal origin** and strains **of human origin** from HC.
The total amount of protein was determined by spectrophotometry using BSA **as a standard**.	The protein content of each well was then determined using the Pierce protein assay, using BSA **as standard**.
We may assume without **any loss of generality** that the quantity "M(ca)" is computable for any M.	For simplicity, and without **loss of generality**, we will assume that …
Without **a loss of generality** we assume that E{|ni|2} = 1.	
This may occur at **an intermediate level**.	This is far more difficult when working **at advanced level**.

6 Quantifiers: *any, some, much, many, much, each, every* etc.

6.1 quantifiers used with countable and uncountable nouns

The table lists words that indicate an indefinite quantity. These are words that you can generally use with countable (1.6) and uncountable nouns (1.8) in a research paper. Note however that the expressions *with a piece of* are not commonly used in research papers.

QUANTIFIER	COUNTABLE (SINGULAR)	COUNTABLE (PLURAL)	UNCOUNTABLE
a / an	a book		a piece of information
a (large / small) amount of		a large amount of books	a small amount of information
a bit / piece of			a piece of information
a few		a few books	
a great deal of		a great deal of books	a great deal of information
a little			a little information
a lot of		a lot of books	a lot of information
a number of		a number of books	
a series of		a series of books	
all		all the books	all the information
any	[see 6.2.4]	any books	any information
each	each book		each piece of information
enough		enough books	enough information
every	every book		every bit of information
few		few books	
little			little information
many		many books	many pieces of information
most		most books	most (of the) information
much			much (of the) information
no	no book	no books	no information

(continued)

A. Wallwork, *English for Academic Research: Grammar, Usage and Style*,
DOI 10.1007/978-1-4614-1593-0_6, © Springer Science+Business Media New York 2013

(continued)

QUANTIFIER	COUNTABLE (SINGULAR)	COUNTABLE (PLURAL)	UNCOUNTABLE
none of		none of the books	none of the information
one	one book		one piece of information
several		several books	
some		some books	some information
the	the book	the books	the information

6.2 *any* versus *some*

The following rules apply to *any* and *some* and derivatives (e.g. *something, anywhere, anyone*)

1. As a general rule *any* is used in negative phrases and *some* in affirmative phrases.
2. *Not … any* = zero, *not … some* = not all.
3. *Any* is used to indicate doubt, we are not sure whether the event will take place or not.
4. If you use *any* in a sentence that contains no negation and which is not covered by Rule 3, then it means 'one thing or person at random from all the individuals in the world'. *some* and *someone* mean one particular thing or person, although exactly what or who is not important.
5. *Any* is used in questions where the answer is not known; *some* is used in questions where the expected answer is affirmative (e.g. in offers and some kinds of requests).

	ANY	SOME
1	This did not give **any** interesting results.	This gave **some** interesting results.
2	We were **not** able to understand **any** of the figures – they were all too complicated and unclear.	We were **not** able to fulfill **some** of the referees requests, specifically the first and last requests.
3	The table shows significant results, if **any**, of each test. Some tests may not have given significant results.	The table shows **some** significant results, in fact …
3	If you need **any** clarifications, then do not hesitate to contact me. I don't know if you require clarifications or not.	I need **some** clarifications with regard to points 3 and 8.
4	**Anyone** can tell you that one plus one equals two.	**Someone** is at the door.
4	**Any** book on the subject will tell you all you need to know.	I read about it in **some** book, but I don't remember which one.
5	Excuse me, do you have **any** idea where the local mosque is?	Would you like **some** wine?

6.3 *any* versus *no*

1. *No one* is preferred to *not … anyone* in formal situations such as research papers.
2. *Without* and *hardly* require *any* rather than *not.*

	YES	NO
1	To the best of our knowledge **no one** has found similar results to these.	To the best of our knowledge there is**n't anyone** who has found …
2	You can do this **without any** problems or at least with **hardly any** problems.	You can do this **without no** problems or at least with **hardly no** problems.

6.4 *a little, a few* vs. *little, few*

1. *A little* (uncountable nouns) and *a few* (plural nouns) indicate a limited quantity of something. They could be replaced by *some.*
2. *Little* (uncountable nouns) and *few* (plural nouns) indicate an extremely low or surprisingly low number. They have a negative sense.

	A LITTLE VS. LITTLE	A FEW VS. A FEW
1	We have **a little** time left, so does anyone else have any questions?	We have **a few** more experiments to do, five or six I think, and then we have finished.
2	**Little** is known about this very rare disease.	**Few** researchers have investigated this complex phenomenon.
	Almost nothing is known.	Maybe only two or three researchers.

6.5 *much, many, a lot of,* and *lots of*

1. *Much* is used with uncountable nouns, and *many* with plural nouns.
2. *Lots of* is considered to be too informal, prefer *a lot of* (which some authors still avoid on the basis that it is not sufficiently formal).
3. *A lot of* is usually replaced by *not much* or *not many* in negative phrases.

	YES	WRONG (*) OR TOO INFORMAL
1	There is not **much information** on this topic.	We do not have **many information**.*
1	We have not made **much progress**.	We have not made **many progresses**.*
1	There have been **many advances** in this technology.	
2	We have **a lot of** data on this issue.	We have **lots of data** on this issue.
3	There are **not many** accessible papers on this subject.	There are **not a lot of** accessible papers on this subject.

6.6 *each* versus *every, every* versus *any*

1. *Each* is used when it is important to underline that you are viewing things as individual items, *every* when these things are seen as a mass.
2. Only *each* can be used before a preposition.
3. Some expressions require *every* and not *each.*
4. Often, there is no real difference between *each* and *every.*
5. *Any* = only one, but it is indifferent which one, *every* = all.

	EACH, ANY	EVERY
1	An acronym is a word in which **each letter** stands for another word.	She is only two years of age and already knows **every letter** in the alphabet.
1	**Each patient** was given a slightly different dosage of the medicine.	**Every patient** in their hospital has medical insurance.
	No patient had the same dosage	All patients
1	**Each volume** deals with a different topic.	I have read **every book** on the topic.
1	**Each individual case** will be analysed separately.	**In every case** death occurred within three months.
		In all cases
1	It is **each applicant's** responsibility to ensure that they provide references.	What **every applicant** should know about the interview process.
2	**Each of** them has a different name.	
	= All of them have different names.	
3		Patients will be examined **every** week / **every** three months / **every** third month.
4	**Each time** we do the experiment something goes wrong.	**Every time** we do the experiment something goes wrong.
5	**Any** element in a set can be used.	**Every** element in this set is important.
	Just one element, it does not matter which one	All the elements

6.7 *no* versus *not*

1. *No + noun* and *not + a / the + noun* are similar in meaning. The form *not + noun* is incorrect (e.g. ~~we have **not reason** to suppose that~~).
2. *No + noun* is often replaced with *not + verb + any + noun*.
3. Adjectives that follow the verb *to be* and which are not associated with a noun are generally preceded by *not.*
4. Use *not* before an adverb.
5. *No longer* can be written more informally as *not … any more / longer.*
6. *No + comparative adjective* (19.1) means that the two things compared are equal; *not + comparative adjective* means that the first thing is not, for example bigger or stronger, than the second thing.

	USAGE WITH 'NO'	USAGE WITH 'NOT'
1	There is **no reason** to suppose that this is due to …	This is **not a good reason** for …
1,2	We **encountered no** problems with the calculations.	We did **not encounter** any problems …
1,3	There are **no unusual species** in this area.	It is **not unusual** to find strange species in this area.
1,4	It is **no surprise** that the cardiovascular system is the first organ system to reach a functional state in an embryo.	**Not surprisingly**, the cardiovascular system is the first organ system to reach a functional state in an embryo.
5	This system is **no longer** used.	This system is **not** used **any more**.
		This system is **not** used **any longer**.
6	Verifying X turns out to be **no easier** than verifying Y.	X is **not easier** to solve than Y.
	X and Y have the same level of difficulty.	Y is probably easier to solve than Z

7 Relative pronouns: *that, which, who, whose*

7.1 *that, which, who, whose*

1. Use *that* and *which* for things, and *who* for people.
2. After a preposition, use *which* (things) and *whom* (people). Note the word order.
3. If you put an adjective after the noun it describes, then this adjective should normally be introduced by *that*, *which* or *who.*
4. *Whose* indicates possession.

	YES	NO
1	Apple's first CEO was Michael Scott, **who** ran the company from 1977 to 1982.	Apple's first CEO was Michael Scott **that** ran the company from 1977 to 1982.
2	I have several mobile phones, many **of which** don't work.	I have several mobile phones, many **of that** don't work.
2	This institute employs many people, most **of whom** are technicians.	This institute employs many people, most **of who** are technicians.
3	I met a student **who is** 25 years old. She wrote a document **which / that is** five pages long.	I met a **student 25** years old. She wrote a **document five** pages long.
4	Professor Shirov, **whose** seminal paper was published in 1996, is professor of …	Professor Shirov, **who's** seminal paper was published in 1996, is professor of …

A. Wallwork, *English for Academic Research: Grammar, Usage and Style*,
DOI 10.1007/978-1-4614-1593-0_7, © Springer Science+Business Media New York 2013

7.2 *that* versus *which* and *who*

1. *That* – when you want to define the preceding noun in order to differentiate it from another noun. The resulting clause is often referred to as a "defining", "identifying" or "restrictive clause". Note: in non-scientific / technical English this rule is often ignored.
2. *Which, who* – to add parenthetical information about the preceding noun. Such parenthetical information is not essential to the sentence – if it was removed the sentence would still make sense. In such cases you are not differentiating the noun but simply giving further details. The subordinate clause in which *who* and *which* occurs is generally separated by commas. The resulting clause is often referred to as a "non-defining clause".
3. *Which, who* – to add additional information at the end of sentence. The resulting clause is often referred to as a "connective relative clause".

	YES	WRONG* OR NOT IN RESEARCH PAPERS
1	Google has many offices. I work for the office **that** is in London.	Google has many offices. I work for the office **which** is in London.
1	I collaborate with the Professor Smith **that** teaches economics, not the Professor Smith **that** teaches sociology.	I collaborate with the Professor Smith **who** teaches economics, not the Professor Smith **who** teaches sociology.
2	Google, **which** is a huge company, receives thousands of CVs every day.	Google, **that** is a huge company, receives thousands of CVs every day.*
2	Professor Jones, **who** lectures in political sciences, is 45 years old.	Professor Jones, **that** lectures in political sciences, is 45 years old.*
3	Google sells a lot of advertising, **which** is one way the company gets its money.	Google sells a lot of advertising, **that** is one way the company gets its money.*
3	I work with Professor Ling, **who** I have known for several years.	I work with Professor Ling, **that** I have known for several years.*

7.3 omission of *that, which* and *who*

This is an area of English grammar that can be very confusing and whose rules are not well defined. The only certain rule is that you can never omit *whose*. If in doubt, the simplest solution is to never omit *that, which* and *who*.

1. You cannot omit *that* in a defining clause when the subject of the phrase is also the subject of the verb. However, you can omit *that* when the subject of the verb is different from the subject of the phrase, and when the verb is in the present continuous.
2. You cannot usually omit *which* or *who* when these are used to introduce the final clause in a connective relative clause (7.2.3).
3. You cannot usually omit *which* or *who* in a non defining clause (7.2.1).
4. There are several exceptions to Rule 3: *who* and *which* can be omitted when attributes, ages, job positions and figures (tables etc.) are mentioned.

	NO OMISSION	OMISSION POSSIBLE
1	The professor **that** wrote the article is giving a presentation.	The professor **[that]** we met yesterday is giving a presentation.
		The professor **[that is]** coming tomorrow won the Nobel Prize.
2	Professor Shirov is giving a presentation on life on Mars, **which** should be very interesting.	
	The presentation on Mars will be given by Professor Shirov, **who** works at IMT.	
3,4	Professor Shirov, **who** is arriving tomorrow and **whose** book was published last year, is giving a presentation on life on Mars.	The committee includes a professor **[who is]** considered to be one of the foremost experts in the field.
3,4	Mars, **which** is millions of miles from Earth, is also known as the red planet.	Shirov's apparatus, **[which is]** shown in Figure 2, is easy to set up.
4		Professor Shirov, **[who is]** aged 52 / **[who was]** born in 1980, is an expert on Mars.
		Professor Shirov, **[who is]** a professor of astrophysics at IMT, warned that ...

7.3 omission of *that, which* and *who* (cont.)

5. *Which* + its related verb are often omitted when giving definitions.
6. You can omit *which* or *who* when the words or phrases in apposition are interchangeable.
7. You can omit *which* or *who* in sentences that would otherwise contain a repetition of *which* or *who* in a very short space.

	NO OMISSION	OMISSION POSSIBLE
5		Gold, **[which is]** a metal commonly used in biochip technologies, was exploited in order to provide an interaction surface.
6		The Thames**, [which is]** England's longest river, is located in London.
		= England's longest river, **[which is]** the Thames, is located in London.
7		Professor Shirov, **who is** an MIT professor **[who was]** awarded the Nobel Prize for physics, warned that …

7.4 avoiding ambiguity by using a relative clause in preference to the *-ing* form

1. Be careful of using the *-ing* form when it is not 100% clear whether the *-ing* form is being used in a restrictive or non restrictive sense (7.2.1).
2. Even where there is no ambiguity, the *-ing* form can be replaced with *that* when the *-ing* form has been used to define the previous noun.
3. The use of *having* tends to be confined to mathematics, physics and computer science. Essentially, you can always use *that has* or *that have.*

	YES	NO
1	Edible **jellyfish, which** belong to the order *Rhizostomeae*, are a popular seafood in Asia.	Edible **jellyfish belonging** to the order *Rhizostomeae* are a popular seafood in Asia.
	All jellyfish are *Rhizostomeae.*	Not clear if the author is referring to all jellyfish or just a subset.
	The / Those edible jellyfish that belong to the order *Rhizostomeae* are a popular seafood in Asia.	
	Only some jellyfish are *Rhizostomeae*	
1	Many authors have performed studies **that compare** X and Y.	Many authors have performed studies **comparing** X and Y.
	The studies compare X an Y.	Not clear whether it was the authors or the studies that made the comparison.
	Many authors have performed studies **by comparing** X and Y.	
	The authors compared X and Y in order to make their study.	
2	These are complexes formed by simple ligands **containing / that contain** a maximum of five coordinating centers.	
3	The null set is the set **having / that has** no elements.	A person **having** no job is called 'unemployed'.
3	Markov processes **having / that have** a countable state space are known as …	Those people **having** no house are known as 'homeless'.

7.5 avoid long and difficult-to-read sentences involving *which*

1. In long sentences, replace the *which* clause by beginning a new sentence. Don't worry about repeating the same key word twice.
2. Consider changing the order of the information to make it easier for the reader to understand.

	YES	NOT RECOMMENDED
1	The CNR is the Italian National Research Council and has many **institutes** where innovative research is carried **out. These institutes** are located in various parts of Italy such as Pisa, Turin and Rome.	The CNR is the Italian National Research Council and has many institutes where innovative research is carried out and **which** are located in various parts of Italy such as Pisa, Turin and Rome.
2	The ethyl acetate phase was dried under a gentle stream of nitrogen, **and** was then re-dissolved with 50 mL of the eluent B.	The ethyl acetate phase, **which** had been dried under a gentle stream of nitrogen, was re-dissolved with 50 mL of the eluent B.

7.6 avoid ambiguity with *which*

1. *Which* generally refers to the noun that it follows. So, in cases of possible ambiguity, avoid using *which*. Instead, split the sentence and repeat the subject. In the 'No' example below, the position of *which* initially seems to refer to *Table 2*. But in fact it refers to *set of common rules*.
2. When the *which* clause could refer to several but not all elements, remove *which* and repeat the specific elements. In the second 'No' example below, *which* could refer to A and B, B and C, or even A, B and C.

	YES	NO
1	Each language is characterized by a set of common rules, as reported in Table 2. **This set** highlights the structure of that particular language.	Each language is characterized by a set of common rules as reported in Table 2 **which** highlights the structure of that particular language.
2	Examples include A, B and C. **A and B** are normally established once a month.	Examples include A, B and C, **which** are normally established once a month.

8 Tenses: present, past, future

8.1 present simple vs present continuous: key rules

1. The **present simple** indicates actions or situations which happen repeatedly, all the time, or at any time, including established scientific facts and findings, theorems, definitions, lemmas, proofs etc.

2. The **present continuous** indicates trends and situations that are evolving now, or programmed future events.

	PRESENT SIMPLE	PRESENT CONTINUOUS
1,2	It **is** well known that in many universities how much you **write** (i.e. the quantity) **is often considered** to be more important than how well you **write** (i.e. the quality).	At the moment we **are writing** a paper on …
1,2	Some maps of the world's oceans **show** the widths of the continental shelves.	The patients **are now showing** signs of fatigue.
1,2	Today a wide range of sensor devices **exist** that **alter** their characteristics in response to a stimulus.	Sensor devices **are currently being developed** that will enable researchers to …
1,2	A nonempty subset *H* of a group *G* **is said** to be a *subgroup* of *G*, if under the product in *G*, *H* itself **forms** a group.	We **are forming** self help groups for those with maritial problems.
1,2	The container **holds** the samples.	The conference **is being held** in July.

A. Wallwork, *English for Academic Research: Grammar, Usage and Style*,
DOI 10.1007/978-1-4614-1593-0_8, © Springer Science+Business Media New York 2013

8.2 present perfect: key rules

1. The active form of the **present perfect** is often used in an Abstract to announce a new finding or some new advance in a particular discipline. This usage tends to be confined to the first line of the Abstract, or immediately after the background information has been given. However, to add further details about this innovation/news, the **present simple** or **past simple** are used.

2. When writing a response to the referees regarding your manuscript, use the **present perfect** to announce what additions and changes you have made. However, when you give more details of these changes and why you made them, use the **past simple**.

	PRESENT PERFECT	OTHER TENSE
1	We **have developed** a new system for converting wind into energy.	Our system **works** by harvesting wind from ... We **implemented** the system in a wind farm in ...
1	Dementia is an increasingly common problem in advanced societies and is known to cause ... We **have discovered** a treatment for dementia.	This treatment **consists** of ... It **was** tested on a sample of 543 patients aged over 80.
2	We **have added** a new figure ... Table 3 **has been** deleted. The Conclusions **have been** completely rewritten.	The figure **was** added because ... Table 3 **was** in fact unnecessary. We **decided** to rewrite the conclusions on the basis of Ref 3's comments.
2	*Ref 3's comment:* I suggest the authors remove Table 2 and combine it with Table 1. *Authors reply:* **Done**.	We **opted** to keep Table 3 because ...

8.2 present perfect: key rules (cont.)

3. **Present perfect** for an action, event, or scenario that began in the past and is still current today. This construction is often used to state the background situation as a basis for current research.
4. If there is no past-present reference, then use the **present simple** to indicate a habitual situation (and the **present continuous** for actions taking place now or in the current period of time).
5. **Present perfect** when *once* and *as soon as* mean 'after something has been done'.

	PRESENT PERFECT	OTHER TENSE
3,4	The sea level **has changed** throughout the Earth's history and will continue to do so.	The sea level **changes** every year.
3,4	**Over the last 60 years** English **has transformed** itself from a predominantly writer-oriented language to a reader-oriented language.	If language **transforms** our thinking, do specific languages **transform** it in different ways?
3,4	**Since** the 1990s / **For the last few decades**, English writers **have published** several thousand books.	English writers **typically publish** their work in English, but more and more **are now publishing** in other languages too.
3,4	**Since 2009** there **have been** many other attempts to establish an international readability index [Wallwork et al, 2009; Smithson 2012], but **until now** no one **has managed** to solve the issue of ...	Establishing an international readability index **represents** a frequent topic in the literature [Wallwork et al, 2009; Smithson 2012]. The essential problem **is** how to solve the issue of ...
4	Once/As soon as the initial tests **have been made**, the research effort will focus on ...	Generally speaking, once you **start** reading the book, you soon **become** addicted.

8.3 present perfect: problem areas

1. **Present perfect** (not **present simple**) to state when, with reference to a present situation, you state how long (in days, years, months etc) this situation has been operative.

   Note that *I have been here for a week* means that I arrived one week ago and I am still here. Instead, *I am here for a week* means that perhaps I arrived today (or recently) and that I will stay seven days. See 14.15 for the difference between *since, for* and *from*.

2. **Present perfect** (not **present simple**) to state that this is the first (second, third, etc) time that something has been done.

3. **Present perfect** (not **past simple**) in the Conclusions to summarize what you have done in the paper - the focus is on the writing and construction of the paper (typical verbs: *present, show, describe, explain, outline*).

	YES	NO
1	We **have used** this system **for many years**.	They are **many years** that we **use** this system. We **use** this system **since many years**.
1	We **have not used** this equipment **for** several months.	We **do not use** this equipment **from** several months.
2	We **have been** here **since last Monday / for nearly a week**.	We **are** here **since last Monday**.
3	**It is the first time** that we **have used** this system.	**It is the first time** that we **use** this system.
3	**This is only the second time** that such a result **has been published** in the literature.	**This is only the second time** that such a result **is published** in the literature.
4	Conclusions: We **have presented** a new methodology for teaching English. We **have shown** that … We **have described** three cases where …	Conclusions: We **presented** a new methodology for teaching English. We **showed** that … We **described** three cases where

8.4 past simple: key rules

1. Use the **past simple** when there a clear time reference (e.g. *in 2011, last month, three years ago*).
2. Use the **past simple** when the action is clearly past.
3. Avoid the **present simple** to describe actions that took place in the past. Instead use the **past simple**, particularly to avoid ambiguity (last example below).

	YES	NO
1	In 2012, Carter **suggested** that complex sentences could also lead to high levels of stress for the reader [25].	In 2012, Carter **has suggested** that complex sentences could also lead to high levels of stress for the reader [25].
1	Smith first **used** this procedure more than a decade ago [24].	Smith **has first used** this procedure more than a decade ago [24].
2	This building technique **was exploited** by the ancient Egyptians for the pyramids.	This building technique **has been exploited** by the ancient Egyptians for the pyramids.
3	Galileo Galilei **was** born in Pisa, Italy in 1564. At the age of eleven, Galileo **was** sent off to study in a Jesuit monastery. After four years, Galileo **announced** to his father that he **wanted** to be a monk.	Galileo Galilei **is** born in Pisa, Italy in 1564. At the age of eleven, Galileo **is** sent off to study in a Jesuit monastery. After four years, Galileo **announces** to his father that he **wants** to be a monk.
3	In 2010 the Social Democrats **challenged** the anti-GMO movement. The fact that this party **was** in favor of genetically modified products **meant** that … It is clear that we are referring only to the situation in 2010. We do not know the Social Democrat's position on GMOs today.	In 2010 the Social Democrats **challenges** the anti-GMO movement. The fact that this party **is** in favor of genetically modified products **means** that … The reader cannot know if the party was only in favor of GMOs in 2010, or if they still are today.

8.5 present simple vs past simple: specific rules (aims and methods)

1. To outline main topics of the research and aims of a project: use the **present simple** in the first sentence to describe the whole paper, use the **past simple** to outline what you did.
2. When describing the aim of the project, use the **present simple** if the project is still ongoing, use the **past simple** if your paper describes a finished project.
3. **Present simple** to describe a procedure (method etc) established by other authors, i.e. to state general principles relevant to the procedure.
4. **Present simple** to refer to your methodology, process or procedure if you are just explaining in general how it works, rather than what you did on one specific occasion.
5. **Past simple** to state what the objectives of your experiments were, what equipment was used, how other methods were adapted, what steps were followed etc.

	PRESENT SIMPLE	PAST SIMPLE
1	This paper **outlines** a methodology for establishing the amount of verbosity in a nation's language.	To establish our verbosity index, we **analysed** five languages. We **classified** these languages in terms of x, y, z. On the basis of these results, we then **calculated** the number of …
2	The aim of this research **is** to …	The aim of the project **was** to …
3	A cloze procedure **is** a technique in which words **are deleted** from a text according to a word-count formula. The passage **is presented** to students, who then …	
4,5	In our procedure the students **are first split** into groups by age and level. This grouping **enables** the teacher to …	The aim of our procedure **was** to find a way for teachers to place students into groups. We **used** GroupSoft (GS Inc, USA) which automatically *places* students into groups. We **adapted** the software by adding an additional step in which students *are* preliminary grouped by age.
4,5	Our methodology **consists** of the following steps: First, we **gather** the data. Second, we **sort** the data by …	In the second experiment we **proceeded** as follows: First, we **gathered** the data. Second, we **sorted** the data by …

8.6 present simple, present perfect and simple past: reference to the literature

1. **Past simple** to refer to the literature when you give the date within the main sentence (i.e. not just in parentheses).
2. **Present perfect** to give past-to-present background information. The **present simple** is possible here, but is much less commonly used.
3. **Present perfect** if the method, technique, procedure etc is the subject of the verb and there is no time reference, **past simple** if the author is the subject of the verb. In such cases there may or may not be a time indication.
4. **Past simple** (or less commonly **present simple)** to report what other authors have *suggested, proposed, claimed, implied, hypothesized, put forward* etc, There cannot be a specific time reference if you choose to use the **present simple** in such cases.

	YES	YES
1	In 2007, Carter **suggested** that women are superior to men [25].	
2	Many authors [3, 6, 8, 12] **have claimed** that there is life on Mars.	Many authors [3, 6, 8, 12] **claim** that there is life on Mars.
3	This method **has been used** to investigate both problems [24].	Smith **used** this method to investigate both problems [24].
3	This procedure **has been exploited** by many authors in order to conduct very diverse investigations.	Smith first **used** this procedure in 1996 [24].
3	In support of such treatment, Griggs **has made** the surprising discovery that …	**Recently**, Griggs **made** the surprising discovery that …
4	For instance in [5] the authors **suggested** that a new strategy could be introduced to …	For instance in [5] the authors **suggest** that a new strategy could be introduced to …

8.7 present simple vs past simple: specific rules (results and discussion)

1. Very important: if you use the **simple present** to report your findings it must be 100% clear to readers that you are talking about YOUR findings and NOT what has been reported in the literature. This rule is extremely important and should never be ignored. See 10.4.2 for possible confusion caused by using the passive.
2. **Present simple** to state what a figure or table shows, highlights, describes, reports etc.
3. **Present simple** to discuss your data and results, and to state the implications of your findings. Typically after *show, explain, highlight, believe, mean, indicate, reveal.*
4. Introductory verbs such as *show, highlight, reveal* can be either in the **present simple** or **past simple**, but to talk about what you *found, discovered, noticed* etc only use the **past simple**.

	PRESENT SIMPLE	PAST SIMPLE
1	We found that green and red **produces** white. This only **seems** to happen when the ratio of green to red **is** 6:1. But when the ratio **is** 4:1, this **produces** yellow. The reader may think that 'seems to happen' and 'produces' refer to what other people have found.	We found that green and red **produced** white. This only **seemed** to happen when the ratio of green to red **was** 6:1. But when the ratio **was** 4:1, yellow **was** produced. The reader is clear that you are only talking about what you found.
2	The results are given in Table 4, which **shows** that ... In addition, Figure 1 **highlights** that X equals Y.	
3	We **believe** that this **means** that our method outperforms all previous methods.	
4	These results **highlight** / **highlighted** the importance of carrying out tests in triplicate.	We **found** that best results were achieved by carrying out tests in triplicate.

8.8 present perfect vs present perfect continuous

1. If a situation has existed for a considerable number of years then the **present perfect** is preferred to **present perfect continuous**, if the situation is more recent then both forms can be used with the **continuous** form indicating that the situation may be temporary.

2. Do not use the **present perfect continuous** for completed actions or when you talk about the number of occasions that something has happened or when you specify a quantity [exception: *years, days, hours, minutes* etc].

	PRESENT PERFECT	PRESENT PERFECT CONTINUOUS
1	For **thousands of years** man **has tried** to give a meaning to life.	For **several years** researchers in this field **have been trying** to understand the ...
2	We have **already** written **three papers** on this topic.	We **have been investigating** this problem for **three years**.

8.9 past continuous and past perfect vs simple past

1. **Past continuous** to indicate a long action that was interrupted by a short action.

2. **Past continuous** to indicate two long actions that took place at the same time, **past simple** for a series of non-simultaneous actions.

3. **Past perfect** to highlight when one action took place earlier than a later action. The **past simple** indicates a series of consecutive actions.

	PAST CONTINUOUS, PAST PERFECT	SIMPLE PAST
1	We encountered a problem while we **were loading** the software.	We **downloaded** the software and **installed** it.
2	While I **was studying** I **was also working** full time.	I **studied** at Ulaanbaatar State University. I then **worked** for Mongolian Railways.
3	Two dogs that **had died** for reasons unrelated to this study were used to characterize the approach.	Unfortunately, one of the cats **died** during the experiments.

8.10 *will*

English has many ways to express the future. However, only *will* and the present continuous (8.1) are usually found in research manuscripts.

1. ***Will*** is used for making predictions.
2. ***Will*** can be used to express a hypothesis referring to one specific case, whereas the **present simple** is used for well-known conditions that are applicable to all cases.
3. In the Conclusions section use ***will*** to outline future work.
4. ***Will*** can be used to talk about future parts of the manuscript.
5. Despite Rule 4, prefer the **present simple**, particularly when outlining the structure of the paper.

	WILL	PRESENT SIMPLE
1	We **predict** that demand **will outweigh** supply, and thus house prices **will rise**.	
2	Note that if the water in the container **reaches** a temperature of 100C it **will** boil and this **will cause** damage to the samples.	It is well known that **if / when** water **reaches** a temperature of 100C it **boils**.
3	Future work **will involve** investigating the reasons for these discrepancies.	
4	We **will see** how relevant this is in the next subsection. We **will** now **explain** how x = y.	
5		This paper **is organised** as follows. Section 1 **gives** a brief overview of the literature. A history of the English language **is presented** in Section 2.
5		This feature **is covered** more in depth in the appendix.

9 Conditional forms: zero, first, second, third

9.1 zero and first conditional

1. **Zero conditional** [*if* + *present* + *present* or *present* + *if* + *present*]: to express general truths, logical conclusions and scientific facts. It means 'every time that' or 'whenever'.
2. **First conditional** [*if* + *present* + *will* or *will* + *if* + *present*]: to talk about real future situations, rather than general truths that are always valid.
3. Both **zero** and **first conditional** forms can be used to express logical conclusions; ***will*** is sometimes preferred when there are various stages in a process and the author is talking about a later stage in the process.
4. The **present perfect** can replace the **present simple** in the *if* clause. In such cases both the **present simple** and ***will*** may be found in the dependent clause.

Note *If* refers to the occurrence of an event that may or may not take place, whereas *when* indicates certainty. Examples: *If the alarm goes off, call the police.* (We don't know whether the alarm will ring or not). *I get up in the morning when the alarm goes off* (My alarm always rings every morning).

	ZERO CONDITIONAL	FIRST CONDITIONAL
1,2	Papers **tend** to be rejected if the English **is** poor.	If we **do not receive** the revised manuscript by the end of this month, we **will be forced** to withdraw your contribution from the special issue.
1,2	If all humans **are** prone to corruption, then politicians **are** more prone than others.	If it **is** true, as many authors contend, that Chinese is set to replace English as the international language, then this **will have** profound effects on …
1,2	if you **wish** to advance your academic career you **have** to publish your research in high impact journals.	If the illness **is** in an advanced stage, then treatment **will have** little effect.
3	This leads to the result that if (1) **is** false then (2) **is** also false.	The second property guarantees that if H **is** true initially, then it **will remain** true while P is being executed.
4	If this period **has elapsed / elapses** before x reaches y, then the system **fails**.	If this period **has elapsed / elapses** before x reaches y, then the system **will fail**.

A. Wallwork, *English for Academic Research: Grammar, Usage and Style*,
DOI 10.1007/978-1-4614-1593-0_9,

9.2 second conditional

1. **Second conditional** [*if* + past simple + *would*]: to express hypothetical situations regarding what would happen if certain features, events, scenarios were possible (which are currently not possible).
2. An alternative form of the **second conditional** is to replace the *if* clause with: *was / were + subject + infinitive*. This is a very formal construction and there is no real need to use it.

	YES	NO
1	If the government **raised** taxes further, this **would have** serious consequences.	If the government **would raise** taxes further, this **would have** serious consequences.
1	**Would** the world **be** different if it **were ruled** by women?	**Would** the world be different if it **would be ruled** by women?
2	**Were citizens to pay** their taxes …	
	Were women to rule the world …	

9.3 other uses of *would*

1. *Would* can be used to make a statement more tentative. It is typically used with *appear, seem* and *suggest.*
2. *Would* is sometimes used to refer to how a past situation later developed in the future. In such cases the simple past could also be used, but not *will.*
3. *Would* + infinitive (not *would have* + past participle) is used in indirect speech to report what someone has said.
4. *Would* can be used to report a habit (a series of repeated actions) in the past particularly in reference to someone's behavior; use of the **past simple** in such situations denotes that such actions may have only taken place only once rather than being repeated.

	YES	NO
1	An intriguing hypothesis concerns the development of bilingualism. **It would seem** that this can be favored when … This **would suggest** that …	
2	This is highlighted by Maria Kazlovic. A mentally troubled woman – she **would** commit suicide two years later – she claimed that …	This is highlighted by Maria Kazlovic. A mentally troubled woman – she **will** commit suicide two years later – she claimed that …
2	His description of this species, which he **would** revise completely in the third and final edition of his book, focused exclusively on …	His description of this species, which he **will** revise completely in the third and final edition of his book, focused exclusively on …
3	The experimenter then told the students that she **would return** later to explain how each problem was solved.	The experimenter then told the students that she **would have returned** later to explain how each problem was solved.
3	In the next session, as soon as he sat down, he said that he **would be** unable to concentrate on whatever I might have to say, because he had just seen a ghost.	In the next session, as soon as he sat down, he said that he **would have been** unable to concentrate on whatever I might have to say, because he had just seen a ghost.
4	In the first session he **showed** no strange behaviors. However at the beginning of each subsequent session he **would stare** at the wall for five minutes, then he **would turn** his head towards me and start speaking at great speed.	In the first session he **would show** no strange behaviors. However at the beginning of each subsequent session he would stare at the wall for five minutes, then he **turned** his head towards me and start speaking at great speed.

9.4 present simple versus *would*

1. When you report a hypothesis, i.e. how you (or another author) imagine something might be, use the **present simple** not *would*. Words and phrases such as *assume, assumption, hypothesize, hypothesis, suggest, argue,* and *according to* indicate to the reader that what you are saying is only a tentative proposal and not necessarily a fact.
2. The most common use of *would* is in a real conditional phrase (9.2), where there is *if* (or when *if* is at least implicit).

	PRESENT	WOULD
1,2	Aardvark's hypothesis suggests that the onset of the disease **is associated** with a sudden increase in blood pressure.	Batteaux suggests that the onset of the disease **would only be associated** with a sudden increase in blood pressure **if** the patient **were** over the age of 50, whereas Aardvark had only hypothesized such an event for younger patients.
1,2	According to these authors, mating early in the morning **is** quite advantageous for small insects, as cool temperatures and a high relative humidity **reduce** the risk of desiccation.	Although we have no data that provides concrete evidence of this, we believe that mating early in the morning **would be** quite advantageous for large insects, as cool temperatures …
1,2	Bakali argues that global warming **is causing** an increase in the possibility for a war to gain access to water.	Bakali also argues that the consequences of such a war **would be** catastrophic. In fact **if wars were started** in order to capture water supplies, the West **would certainly be forced** to intervene and this **would lead** to …

9.5 third conditional

1. **Third conditional** [*if* + *past perfect* + *would have* + *past participle* or *would have* + *past participle* + *of* + *past perfect*]: to express how things might have been if something had (not) happened.
2. *Would have* + *past participle* can also be used in implied conditions.
3. *Would have* + *past participle* is not used to report what other people have said, unless what they originally said contains a third conditional.
4. An alternative form of the **third conditional** is to replace the *if* clause with: *had* + subject + *past participle*. This is a very formal construction and there is no real need to use it.

	YES	NO
1	This mistake **would not have been made** if the authors **had been** more careful.	This mistake would not have been made if the authors **would have been** more careful.
1	What **would have happened** if the central bank **had not intervened**?	What would have happened if the central bank **would not have intervened**?
2	This work **would not have been** possible without the help of the following people:	This work **was** not possible without the help of the following people:
	= If the following people had not helped me.	
3	One juror said that she **would have done** exactly what the defendants had done if she **had been** in their position.	One migrant said that he **would have only liked** to live in a country where everyone followed regulations and valued fairness.
	The original words of the juror were: *I would have done exactly what the defendants did if I had been in their position.*	The original words of the migrant were: *I would only like to live in a country where everyone follows regulations and values fairness.* Thus the correct version of the above is: *he would only like to live.*
4	**Had the physician known** the true nature of the patient's condition, she would have never prescribed such a high dose.	
	= If the physician had known …	

10 Passive versus active: impersonal versus personal forms

Form the passive as highlighted in the examples:

Active: They *built* a new road. (past simple)

Passive: A new road *was built*. (past simple of *to be* + past participle)

Active: They *are building* a new road. (present continuous)

Passive: A new road *is being built*. (present continuous of *to be* + past participle)

Most books on good writing advocate preferring the active form to the passive form. Also, software applications that automatically check a text for grammar and style, will usually highlight any usages of the passive and recommend using the active as an alternative.

However, in research mansucripts, the passive is often a much better option (see 10.1 and 10.2).

A. Wallwork, *English for Academic Research: Grammar, Usage and Style*,
DOI 10.1007/978-1-4614-1593-0_10,

10.1 main uses of passive

The passive is generally used in preference to the active in all the cases below. The active equivalent might be preferential when giving oral presentations or in other more informal contexts:

1. To describe processes. In such cases the main interest is not in who or what carried out the actions; the most important item is the subject of the sentence. Typically this is found in the Methods section. Only use *we* in the Methods if it is not clear who carried out the action.
2. When making general references to the literature or to what is happening in the world in general.
3. When it is unnecessary, difficult, or impossible to identify the originator of the action.
4. To report what is commonly believed to be true.
5. To report formal decisions or to make announcements.

	PASSIVE	ACTIVE
1	The rust **was removed** by acid-treatment.	We **removed** the rust by acid-treatment.
1	An aerosol solution **was added** to make the flame front visible.	We **added** an aerosol solution to make the flame front visible.
2	Several attempts **have been made** to explain this phenomenon [17, 24, 33].	Several researchers **have attempted** to explain this phenomenon [17, 24, 33].
2	Much progress **is being made** in the field of telecommunications.	They **are making** much progress in the field of telecommunications.
3	The surface of the steel piping **was fractured**.	Something **had caused** the steel piping **to fracture.**
3	A large sum of money **was recently donated** to the hospital.	Someone recently **donated** a large sum of money to the hospital.
4	This drug **is known** to have serious side effects.	Serious side effects **typically arise** from the use of this drug.
5	The law **is expected to be passed** next year.	It is likely that the government **will pass** this law next year.

10.2 passive better than active: more examples

Consider using the passive to avoid:

1. An excessive use of *we* and *us* (15.8), but prefer an active form if possible.
2. Using the impersonal form *one*. The use of *one* has become quite archaic.
3. Sequences of nouns.

Note:

4. The passive is generally used with verbs such as *install, upload* and *download.*

	YES	OK (1–3), WRONG (4)
1	An example of this effect **is shown** in Figure 4. = Figure 4 **shows** an example of this effect.	We **show** an example of this effect in Figure 4.
1	The example **can be strengthened** by means of the circuit in Fig. 3b.	Let us **strengthen** the example by means of the circuit in Fig. 3b.
2	On the other hand the other case of a branch **is only obtained** at the TTC input.	On the other hand one **obtains** the other meaning of a branch only at the TTC input.
3	Costs **can be further reduced** since the components **can be placed** in arbitrary positions in the memory space.	Further **reductions** in costs follow from the **possibility to place** the components in arbitrary positions of the memory space.
4	The system **is installed** automatically.	The system **installs** automatically.
4	Files **are downloaded** directly from source.	Files **download** directly from source.

10.3 active better than passive

1. An active sentence helps to the reader to understand exactly who is the agent (in our case, the author / researcher) of an action. Thus, if your journal permits the use of *we*, then use *we* to avoid any confusion about whether you or another author performed a certain action (10.4).
2. Active sentences do not necessarily have to be personal. Use the active form if this helps to shift the verb nearer to its subject.
3. Some passive constructions sound awkward or wrong in English, particularly with the verbs *to aim* and *to focus.*

	YES	NOT RECOMMENDED
1	We **compared** our results with those of Alvarez.	The results **were compared** with those of Alvarez. Possibly ambiguous, but fine if it is clear from the context who did what
2	The following section **outlines** the state of the art in cybertronics.	In the following section the state of the art in cybertronics **is outlined**.
2	Figure 1 **shows** the relevant trends.	The relevant trends **are shown** in Figure 1.
2	The system **supports**: x, y and z.	The following features **are supported** by the system: x, y and z.
3	The main aim of this project **is** to develop an alternative to the Internet.	This project **is mainly aimed** at developing an alternative to the Internet.
3	This paper **focuses** on the best way to control the activities of potentially rogue traders.	This paper **is focused** on the best way to control the activities of potentially rogue traders.

10.4 ambiguity with passive

Some journals insist that you do not use the personal pronoun *we*. This means that instead of writing *we did x* (active), you have to write *x was done* (passive). Unfortunately, the passive form does not tell the reader with 100% certainty who performed the action.

1. If you use the passive to talk about something which is commonly referred to in the literature, then it will help the reader if you use a word or expression that indicates that this is common knowledge.

	YES	POSSIBLY AMBIGUOUS
1	Children are conditioned by their parents [1, 7, 9]. Thus **it is generally assumed** that children in orphanages will …	Children are conditioned by their parents [1, 7, 9]. Thus **it is assumed** that children in orphanages will …
	generally indicates that this is an assumption made in the literature and not specifically by the authors of this paper	It is impossible to understand who has made or is making the assumption.
1	Children are conditioned by their parents [1, 7, 9]. **It is well known that** children who have been abandoned by their parents will …	
	it is well known clarifies that this is not just the author's viewpoint.	

10.4 ambiguity with passive (cont.)

If you are talking about the literature and you use the passive both to refer to your own work and that in the literature, then the reader will have difficulty distinguishing between the two. There are various devices that are essential to avoid such confusion:

2. Use the names of authors preferably within the main sentence and use the active form. The problem with only using the reference without the name of the author, is that the reader is forced to check to see in the bibliography whether the reference refers to you or to another author.
3. Although some journals dislike *we*, they don't seem to have problems with *our*! So one good way to avoid possible misunderstanding is to use expressions such as *our results show, in our work, in our study.* Using such expressions is vital when you are constantly switching from talking about the literature to talking about your work.
4. Be careful when using expressions such as *in a previous work* – it must be very clear that you are talking about your own previous work, rather than the previous work of an author you have just mentioned.

	YES	POSSIBLY AMBIGUOUS
2	**Peters found** that children perform such tasks better than adults [34].	**It was found** that children perform such tasks better than adults [34].
3	These features are generally characteristic of this species [Smith 2010, Carsten 2013]. However, **in our study, it was found** that they are also characteristic of some completely unrelated species.	These features are generally characteristic of this species [Smith 2010, Carsten 2013]. **However, it was found** that they are also characteristic of some completely unrelated species.
4	Ying et al. noted that red is most people's favorite color. However, **in a previous work carried out by our group, it was noted** that green was …	Ying et al. noted that red is most people's favorite color. However, **in a previous work it was noted** that green was …

11 Imperative, infinitive versus gerund (*-ing* form)

11.1 imperative

The imperative is formed with the infinitive without *to*. It is used in manuscripts in order to:

1. Remind the reader of certain information, or bring attention to certain facts.
2. Give hypotheses.
3. Refer the reader to other sections in the paper or external documents.

	YES	YES
1	**Recall** that x = 1.	**Note** that the values of x may vary.
2	[Let us] **Suppose** that x = 1.	**Let** x be equal to 1.
3	This is of great importance (**see** below).	**See** Smith's paper [23] for details.

A. Wallwork, *English for Academic Research: Grammar, Usage and Style*,
DOI 10.1007/978-1-4614-1593-0_11,

11.2 infinitive

1. Use the infinitive when you talk about the aim / purpose of an action, or how to carry something out.
2. Do not precede the infinitive with *for.*
3. The negative infinitive is: *in order not to, so as not to.*
4. Most adjectives (including superlatives) are followed by the infinitive.
5. When a quantifier (e.g. *enough, too much, too many, too little, too few*) is followed by a *noun* + *verb* construction, the verb is in the infinitive.

	YES	NO
1, 2	**To make** extra money, he designs and develops software.	**For to make** extra money, he designs and develops software.
		For making extra money, he designs and develops software.
2	I need money **to buy** a house	I need money **for buying** a house
3	**In order not to lose** data, make back-ups regularly.	**For not losing / For don't / To don't lose** data, make back-ups regularly.
4	It is **straightforward to verify** that x=y.	It is **straightforward verifying** that x=y.
5	It has been claimed that five users is **enough to catch** 85% of the problems on the vast majority of websites.	It has been claimed that five users is **enough for catching** 85% of the problems on the vast majority of websites.
5	There are too few studies with too **few patients to determine** which is the best drug.	There are too few studies with **too few patients for determining** which is the best drug.

11.3 in order to

1. There is a tendency to use *in order to* in more formal situations.
2. Use *in order to* if one infinitive is immediately followed by another.
3. The use of *in order to* rather than simply *to* is often optional, and sometimes redundant.
4. *In order* is not necessary when the focus is on the activity rather than the purpose.

	IN ORDER TO	TO
1	**In order to drive** a car, a license must be obtained.	**To drive** a car you need a license.
2	Having an English dictionary is very important, in fact a dictionary is vital **in order to be able to distinguish** between different meanings of the same word.	It is vital **to learn** English if the desired outcome is **to be** successful.
2	If a scientist feels it necessary, therefore, **to publish in English in order to** reach a worldwide audience, does this mean that …?	It is necessary **to publish in English if you wish to reach** a wider audience.
3	**[In order] to learn** English it helps to have a good teacher.	**To learn** English it helps to have a good teacher.
3	Our librarian will consult the library collection **[in order] to see** if we already have these books.	Our librarian will consult the library collection **to see** if we already have these books.
4	**[In order] to teach** English, candidates are required to have a certificate.	There is now a program of retraining Russian teachers **to teach** English.

11.4 passive infinitive

The passive infinitive is formed by the verb *to be* + past participle.

1. It is used when the verb that follows the noun is not the subject of that noun (i.e. when something else is responsible for the action).

2. In some cases both forms are possible. In the example, the normal infinitive possibly indicates that the reader is expected to do the tasks, whereas the passive leaves this more open.

	NORMAL INFINITIVE	PASSIVE INFINITIVE
1	This enables us **to calculate** the ratio.	This enabled the ratio **to be calculated**.
1	We still need **to identify** the variants that influence these traits.	The variants that influence these traits still need **to be identified**.
1	In order **to see** these readings, we shifted the corresponding points horizontally and connected by straight lines.	To enable these readings **to be seen** separately, the corresponding points were shifted horizontally.
1	We remained after the presentation **to see** Professor Yi's experiments.	It remains **to be seen** whether the government will actually implement this policy.
2	Below is a list of tasks **to do** next week.	Below is a list of tasks **to be done** next week.

11.5 perfect infinitive

The perfect infinitive is formed by *to* + *have* + past participle.

1. It is used when it is important to underline that something happened in the past, rather than being true on all occasions (normal infinitive).

	NORMAL INFINITIVE	PERFECT INFINITIVE
1	Our clustering algorithm seems **to perform** very well with whatever kind of data it has to deal with.	In the last experiment, the clustering algorithm seems **to have performed** very well, with just a few individuals falling outside the obvious clusters.
1	Malaria is estimated **to cause** almost one in five deaths in sub-Saharan Africa.	This disease was estimated **to have caused** or contributed to death in 122 of 51,645 of the patients analysed.
1	Around 10,000 people claim **to see** UFOs on a regular basis.	Around 100,000 people claim **to have seen** a UFO last year.

11.6 gerund *(−ing* form): usage

The gerund (also known as the *-ing* form) is formed by adding *-ing* to the bare infinitive form (e.g. *study + ing = studying*). The negative is formed by putting *not* in front (e.g. *not studying, not working*). Use the gerund:

1. When the verb is the subject of the sentence.
2. After a preposition, adverb or conjunction.
3. Do not use the gerund when you are talking about an aim, objective or target. Instead, use the infinitive (11.2).

	YES	NO
1	**Developing** software is their core business.	**To develop** software is their core business.
2	Before **starting** up the PC, make sure it is plugged in.	Before **to start** up the PC, make sure it is plugged in.
2	When **transferring** the samples, ensure that the recipient is clean.	When **transfer** the samples, ensure that the recipient is clean.
2	The contents may be displaced while **being** transferred.	The contents may be displaced while **to be** transferred.
3	Our aim is **to investigate** the use of X.	Our aim **is investigating** the use of X.
	= **Investigating** the use of X is our aim.	
3	The target **was to identify** those elements that require X.	The target **was identifying** those elements that …
	= **Identifying** those elements that require X was the target.	

11.7 *by* versus *thus* + gerund to avoid ambiguity

1. Use the gerund at the beginning of the sentence when it is the subject of the main verb.
2. When something else is the subject of the main verb, then the gerund must be preceded with *by* or replaced with an *if* clause.
3. Use *thus* plus the gerund to indicate the consequence of doing something.
4. Using *by* instead of *thus*, and vice versa, can completely change the meaning of the sentence.

	YES	NO
1	**Learning** English **will help you** to pass the exam.	
2	**By learning** English **you** will pass the exam. = If you learn English you will pass the exam.	**Learning** English **you** will pass the exam.
2	**By clicking** on the mouse you can open the window = If you click …	**Clicking** on the mouse you can open the window.
3	We learn English **thus enabling** us to communicate with our international colleagues. = We learn English and thus we can communicate … = We learn English and this means we can communicate …	We learn English **enabling** us to communicate with our international colleagues.
3	The introduction of the euro led to a rise in prices **thus causing** inflation. = The introduction of the euro led to a rise in prices *and this caused inflation.*	The introduction of the euro led to a rise in prices **causing** inflation.
4	This improves performance **by keeping** customers satisfied. = Performance improves when customers are satisfied.	This improves performance **keeping** customers satisfied.
4	This improves performance **thus keeping** customers satisfied. = If performance improves then customers will be satisfied.	This improves performance **keeping** customers satisfied.

11.8 other sources of ambiguity with the gerund

1. If you begin a sentence with the gerund, the reader may not be clear who or what this gerund refers to. Solution: rearrange the sentence using a *subject + verb* construction.
2. When the gerund appears in the second part of a sentence it may not be clear if it refers to the subject or the object of the verb in the first part. Solution: Use *that* or *because* (*since, as* etc.) to clarify.
3. If you are simply giving additional information, use *and* (this is not a rule but facilitates the reader's understanding).

	YES	NO
1	Since **the frequency spectrum is equal** for all the examined transients, the curves have the same shape and differ only in amplitude.	**Being equal** for all the examined transients **the frequency spectrum**, the curves have the same shape and differ only in the amplitude.
1	If **the status is set** to OFF, users will not be able to operate the machine.	**Setting the status** to OFF, users will not be able to operate the machine.
1	After **the gels had been washed** to remove impurities, they were incubated for 90 min.	After **washing** to remove impurities, **the gels** were incubated for 90 min.
2	Professor Yang only teaches students **that have** a good level of English.	Professor Yang only teaches students **having** a good level of English.
	It is clear that it is the students who have the good level of English.	Who has good English – the students or Yang?
2	Suzi teaches students **since / because** she has a passion for teaching.	Suzi teaches students **having** a passion for teaching. Who has passion – the students or Suzi?
3	This document gives an overview of X **and throws** light on particular aspects.	This document gives an overview of X, **throwing** light on particular aspects.

11.9 replacing an ambiguous gerund with *that* or *which*, or with a rearranged phrase

1. The gerund can be ambiguous when the reader is not sure whether you are using it in a restrictive or non restrictive sense (7.2); use *that* or *which* to clarify.
2. In some cases the best solution is to rearrange the sentence.

	AMBIGUOUS	NOT AMBIGUOUS	NOT AMBIGUOUS
1	Phenolic resin components (PRCs) **occurring** on the surfaces of plant organs have been frequently used, particularly in medicines.	Phenolic resin components (PRCs) **that occur** on the surfaces of plant organs have been frequently used, particularly in medicines.	Phenolic resin components (PRCs), **which occur** on the surfaces of plant organs, have been frequently used, particularly in medicines.
	Does this mean all or just some PRCs?	Not all PRCs occur on plant organs	All PRCs occur on plant organs, this is just additional information
2	A horizontal force is applied to one cylinder at a constant rate **measuring** the corresponding displacement.	A horizontal force is applied to one cylinder at a constant rate. **This rate measures** the corresponding displacement.	A horizontal force is applied to one cylinder at a constant rate. **This force** is then used to measure the corresponding displacement.

11.10 verbs that express purpose or appearance + infinitive

1. Verbs that express purpose / objective: *afford*, *attempt*, *choose*, *compel, convince, decide, encourage, force, hope, intend, invite, learn, manage, neglect, oblige, offer, order, plan, persuade, prefer, promise, propose, refuse, remember, study, teach, try, want, warn, wish, would like.*
2. Verbs that express appearance*: appear, seem.*

	YES	NO
1	We are **planning to have** a meeting next week.	We are **planning having** a meeting next week.
1	I **write to inform** you that your invoice has now been processed.	I **write informing** you that your invoice has now been processed.
2	This **seems / appears to** be the best solution.	This **seems / appears being** the best solution.

11.11 verbs that require an accusative construction (i.e. person / thing + infinitive)

Some verbs when used in the active require a direct object before the infinitive. There are three typical constructions:

1. X allows Y to do Z.
2. X allows Y to be done.
3. X is allowed to do Y.

Some common verbs that follow all three rules (*means they only follow Rules 1 and 2): *advise, ask, encourage, force, oblige, offer*, promise*, prefer*, request, want*, wish*, would like*; allow, enable, permit, predict, expect, forecast*

	YES	NO
1	The build-up of large water masses against the shore **forces** the water **to move** seaward as an undertow.	The build-up of large water masses against the shore **forces the water moving** seaward as an undertow.
1	A passport **permits** the holder **to travel** across national borders.	A passport **permits traveling / to travel** across national borders.
1	The referees **want / have asked / have requested** us to make various changes.	The referees **want / have asked / have requested we** make various changes.
1	I **would like** you **to make** the following changes:	I **would like that you make** the following changes:
2	This software **allows** tasks **to be carried out** more quickly.	This software **allows to carry out** tasks more quickly.
2	The editors **expect the changes to be made** before the end of the month.	The editors **expect that the changes are made** before the end of the month.
3	Ph.D. students **are encouraged to present** posters at the conference.	
3	With this password **users are enabled to use** the system.	With this password **users enable to use** the system.

11.12 active and passive form: with and without infinitive

1. The infinitive is used after the passive form (but not after the active) with the following verbs: *assume, believe, hypothesize, imagine, suppose, think*. These verbs all express some kind of opinion or reasoning.
2. When the verbs listed in Rule 1 are used in the active, a different construction is required (*that* + noun + verb in active form).
3. If the subject of the passive form is *it*, then the same construction as in Rule 2 is required.

	PASSIVE	ACTIVE	NO
1,2	The value of x **is assumed to be** equal to 1.	We **assume** that the value of x **is** equal to 1.	We **assume** the value of x **to be** equal to 1.
1,2	This tree **was believed to have** supernatural powers.	They **believed** that this tree **had** supernatural powers.	They believe **to have found** the answer.
3,1	It **was thought** that the answer was known.	They **thought** they knew the answer.	They **thought** to know the answer.
3,1	It **was assumed** that the problem had been resolved.	We **assumed** that **we had** already resolved this problem.	We **assumed to have** already resolved this problem.

11.13 active form: verbs not used with the infinitive

The following verbs are not followed by the infinitive in the active form: *believe, realize, think*. Instead use this formula: *verb + (that) + pronoun + suitable tense*

YES	NO
We **believe** **(that) we are** the first to have revealed this discrepancy.	We **believe to be** the first to have revealed this discrepancy.
We **realized** **(that) we had** this problem only a month ago.	We **realized to have** this problem only a month ago.
She **thought** **(that) she was** right.	She **thought to be** right.

11.14 *let* and *make*

1. *To* is not used after *make* (in active sentences) and *let*.
2. *To* is used after *make* in passive sentences.
3. Do not use *let's* (i.e. the contracted form of *let us*). *let's* is considered too informal.
4. *Let* is often used when giving preliminaries. The verb after *let* is in the infinitive form (which is actually the present subjunctive form).

	YES	NO
1	The engine **makes the wheels go** round.	The engine **makes the wheels to go** round.
1	Please **let me know** as soon as possible.	Please **let me to know** as soon as possible.
2	He **was made to write** the paper by his professor.	He **was made write** the paper by his professor.
3	**Let us** now look at Equation 5.	**Let's** now look at Equation 5.
4	**Let X be** a compact convex set in a topological vector space Y.	**Let X to be / Let X is** a compact convex set in a topological vector space Y.

11.15 verbs + gerund, *recommend, suggest*

1. Verbs that are followed by some kind of activity or course of action tend to take the gerund. The following are just some examples: *avoid, carry on, consider, contemplate, delay, entail, finish, imagine, imply, mean, miss, postpone, recommend, risk, suggest.*
2. *Prevent* and *stop* are followed by an *object + gerund* construction.
3. Use the gerund after *to* in these verbs: *be dedicated to, be devoted to, be an aid to, look forward to, contribute to, object to.*
4. When a recommendation or suggestion is made to a third party, then use the following construction: *recommend / suggest* that someone [should] do (infinitive form) something.

	YES	NO
1	The survey also showed that 88% of these graduates were satisfied with their programs of study and would **recommend studying** in Scotland.	The survey also showed that 88% of these graduates were satisfied with their programs of study and would **recommend to study** in Scotland.
1	Tagawaki et al. have **suggested doing** this in reverse order.	Tagawaki et al. have **suggested to do** this in reverse order.
1	This **entails carrying** out further tests.	This **entails to carry out** further tests.
1	We have **finished writing** the first draft.	We have **finished to write** the first draft.
2	Does parental disapproval **prevent teenagers from drinking** alcohol?	Does parental disapproval **prevent teenagers to drink** alcohol?
2	How do we **stop doctors [from] overprescribing** antibiotics?	How do we **stop doctors to overprescribe** antibiotics?
		How do **we stop that doctors overprescribe** antibiotics?
3	Most of this section is **devoted to reviewing** the literature.	Most of this section is **devoted to review** the literature.
3	I look forward **to hearing** from you.	I look forward **to hear** from you.
4	The referees **recommend / suggest that you / he [should] reorganize** the structure of your / his paper.	The referees **recommend / suggest you / him to reorganize** the structure of your / his paper.
4	We **recommend / suggest that** policy changes in this direction **[should] be made.**	We **recommend / suggest to make** policy changes in this direction.

11.16 verbs that take both infinitive and gerund

Sometimes the same verb can take either the infinitive or the gerund, depending on its meaning:

1. They take the infinitive when the focus is on the purpose or objective.
2. They take the gerund when the focus is on the activity.
3. *Start* and *begin* can be followed by either the infinitive or gerund with no apparent change in meaning. However if *start* and *begin* are in a continuous form (e.g. *is starting, was beginning*), then they are followed by the infinitive.
4. After *it is used* you can either use the infinitive or *for* + gerund.

	INFINITIVE	GERUND
1,2	The experiments on the animals **were stopped in order to avoid** any further protests by activists.	We **stopped doing** the experiments to avoid protests by animal activists.
1,2	Please **remember to include** your biography with your manuscript.	The patient **remembered dreaming** about his mother the night before.
1,2	We **regret to inform** you that we cannot accept your proposal.	I **regretted not accepting** the job proposal.
1,2	We would **like to emphasize** that …	I **like playing** all kinds of sports.
3	She **teaches young children to dance** in her spare time.	She **teaches dancing** in her spare time.
3	I am **starting to learn** Spanish.	I have **started to learn / learning** Spanish.
4	A pen **is used to write** with.	A pen **is used for writing** with.

12 Modal verbs: *can, may, could, should, must* etc.

12.1 present and future ability and possibility: *can* versus *may*

1. *Can* indicates a characteristic behavior. When certain conditions are met or desired, *can* indicates that things are possible but do not necessarily happen.

2. *May* indicates only the potential for something to happen. It indicates uncertainty and is thus used to make hypotheses, to speculate about the future, or to talk about probability.

	CAN	MAY
1,2	Bilinguals are people that **can** speak two languages.	Bilinguals **may** sometimes have learning difficulties when very young.
1,2	Government cuts in education funding **can** have devastating effects on research (Ref. 12–28).	In the next decade such government cuts **may** lead to the closure of several universities.
1,2	This situation **can** be [= This situation is] quite dangerous when hydrogen is present in the chamber. Such dangers **can** be mitigated by properly designing the compartments.	It **may** be dangerous to speculate about the possibilities of this actually happening as so many factors are involved.
1,2	It **can** rain [= It rains] a lot during a monsoon, up to 20 cm of rain at one time.	It usually rains a lot during a monsoon, but this year it **may** rain less as a result of global warming.
1,2	From this perspective, the costs of low short-term interest rates **can** be seen largely as adjustment costs.	Interest rates **may** go up again in the near future.
3,2	I **can** see [= I will see] you tomorrow – what time shall we meet?	I **may** be here tomorrow, but I am not 100% sure.

A. Wallwork, *English for Academic Research: Grammar, Usage and Style*,
DOI 10.1007/978-1-4614-1593-0_12, © Springer Science+Business Media New York 2013

12.1 present and future ability and possibility: *can* versus *may* (cont.)

3. *Can* indicates certainty regarding the future.
4. *May have + past participle* is used to indicate a deduction made about a past event. Note: the form *can have + past participle* does not exist.
5. Sometimes there is very little difference in meaning when *can* and *may* are used in the affirmative form, though *can* indicates greater certainty and is therefore preferred in definitions (last example below).

	CAN	MAY
4		Our sample was only small. Clearly, this **may have affected** the results.
5	In our view, having two systems **can / may** be a more reliable way for dealing with this problem.	In our view, having two systems **can / may** be a more reliable way for dealing with this problem.
5	Dogs **can / may** eat up to 5 kg of food per day, as can be seen in Table 4.	Dogs **can / may** eat up to 5 kg of food per day, as can be seen in Table 4.
5	A university **can** be defined as a place of advanced learning.	A university **may** be defined as a place of advanced learning.

12.2 impossibility and possibility: *cannot* versus *may not*

1. *Cannot* indicates impossibility (i.e. a certain event or scenario is not possible).
2. *May not* indicates there is a possibility that something will not happen (i.e. a certain event or scenario is not likely).
3. *Cannot have* + past participle indicates a deduction regarding the impossibility of a past event.
4. *May (not) have* + past participle is used to speculate about the past, particularly in the Discussion; *might have* and *could have* can also be used in the same way. Note: the form *can have* + past participle does not exist.

	CANNOT	MAY NOT
1,2	I apologize, but I **cannot** come to the meeting as I will be in Hong Kong.	I **may not** be able to come to the meeting tomorrow – is it alright if I let you know later today?
1,2	It is well known that most North Americans and Britons **cannot** speak any foreign languages.	Professor Smith is English so he **may not** speak any foreign languages.
3,4	Shakespeare was not born until 1564 so this work (dated 1560) **cannot** have been written by him.	Although our sample was only small, this **may not** have affected the results because the sample was, in any case, very representative.

12.3 ability: *can, could* versus *be able to, manage, succeed*

1. *Can* is used to talk about a future ability provided that the decision is being made now. Note: *will can* is incorrect.
2. In cases where *can* is not possible to talk about a future event, a form of *be able to* is generally used.
3. *Could* in the affirmative indicates a habitual past ability, i.e. something that someone or something was able to do regularly; *was able to* can also be used in this context. Note: like all modal verbs *could* requires the infinitive without *to.*
4. When describing an ability to do something on one particular past occasion, *could* is never used in the affirmative and interrogative forms. In both these cases, use a form of *to be able to, to succeed in* or *to manage.*
5. In order to avoid confusion with the conditional, to talk about past inabilities it is better to replace *could not* with *did not manage to, did not succeed in,* or *was / were not able to.*
6. *To be able to* replaces *can* in all other tenses and forms.

	YES	NO
1	I **can** finish the paper by tomorrow.	I **will can** finish the paper by tomorrow.
2	I **will be able to** speak better English when I have finished this course.	I **can** speak better English when I have finished this course.
3	The patient **could / was able to** walk at the age of six months.	The patient **could to** walk at the age of six months.
4	I **managed / was able to** finish the manuscript on time. I **succeeded** in finishing the …	I **could** finish the manuscript on time.
5	They **didn't manage / were unable to** do it. They **didn't succeed** in doing it.	They **couldn't** do it. potentially ambiguous
6	We **would have been able** to obtain better results if …	We **would have been can** obtain better results if …
6	In order **to be able** to make this calculation, the following are required:	In order **to can** make this calculation, the following are required:

12.4 deductions and speculations about the present: *must, cannot, should*

1. *Must* is used for drawing logical conclusions in the affirmative form; *have to* is not generally used in such contexts.
2. *Cannot* is used for drawing logical conclusions in the negative form.
3. *Should* indicates what is likely (but not certain) to happen.

	YES	NO
1	If X=1 and Y=2, then X+Y **must** equal three.	If X=1 and Y=2, then X+Y **has to** equal three.
2	If X=1 and Y=2, then X+Y **cannot** equal five.	If X=1 and Y=2, then X+Y **must not** equal five.
3	If the two substances are mixed together, they **should** go red. However, occasionally the mixture is brown.	If the two substances are mixed together they **must** go red. However, occasionally the mixture is brown.

12.5 deductions and speculations: *could, might (not)*

1. *Could* is often used to suggest a possible course of action.
2. *Might* indicates a possible reaction to or consequence of a course of action – but there is no certainty that this reaction or consequence will take place.
3. The difference between *could* and *might* is occasionally very subtle – *could* has the sense of certainty, *might* of uncertainty (this may or may not happen).
4. Sometimes *could* and *might* can be used interchangeably.
5. *Could not* is not used to make speculations, instead *cannot* is used; *might not* means that there is a possibility that something is not true.

	COULD	MIGHT
1,2	Future research **could** be directed towards elucidating this pathology.	Such research **might** then reveal the true causes of this pathology.
1,2	One solution **could** be to get parents and children to swap roles for a day.	What if parents and children swapped roles for a day? How **might** they behave differently?
1,2	We **could**, of course, increase the use of transgenic crops without thinking too much about the consequences.	We show that major problems **might** result from excessive use of transgenic crops over time.
1,2	If we had more energy then we **could** certainly increase production.	We **might** be able to increase production, but only if the following set of requirements were all complied with.
3	These factors **could [=can]** be interpreted as being indicative of …	Unfortunately, the referees **might [=may]** interpret our findings as being indicative of …
4	The temperature then rises dramatically. This effect **could / might** be due to … and this **could / might** explain why …	The history of the world **could / might** be categorized as a series of random events.
5	This **cannot** be the reason why the first two experiments gave very different results. There must be another reason …	This **might (may) not** be the reason why the first two experiments gave very different results. There is a possibility that there are other explanations …

12.6 present obligations: *must, must not, have to, need*

must and *must not* are not frequently found in papers, but are often found in specifications or instruction manuals. Forms of *to have to* are rare in papers.

1. *Must* means that something is an absolute requirement given by a specific authority. Note: *to have to* is not generally used in such circumstances.
2. *To have to* is used to report what an external authority has decided.
3. *Must not* means that something has been prohibited by an authority.
4. *Do / does not have to* means that something is not mandatory. The forms *hasn't to*, *haven't to* and *hadn't to* are incorrect.
5. *Need* indicates necessity and may be used to make a recommendation.
6. *Do / does not need* mean approximately the same as *do / does not have to*. Note: although there is a distinction between *do not need* and *needn't* it is not relevant for research papers.

	MUST	HAVE TO, NEED
1,2	Helmets **must** be worn on the building site at all times.	You **have to** wear a helmet on the building site at all times.
1,2	The form **must** be filled out and signed by the applicant.	I think we **have to** fill out the form and then sign it.
	Please ensure that the form is filled out by the applicant.	
3,4	Authors **must not** copy the text of other authors.	As a Ph.D. student, I have to write a dissertation in my third year. However, I **don't have to** write it in English – I also have the option of writing it in my own language.
5		This area **needs** further investigation.
6		We **don't need to** do it tomorrow, we can do it next week if you want.
		= We **don't have to** do it tomorrow, we can do it next week if you want.

12.7 past obligation: *should have* + past participle, *had to, was supposed to*

1. *Must* has no past form. When you refer to a past obligation that was fulfilled (i.e. you did what you were obliged to do), use *had to, didn't have to.*
2. *Was / were supposed to* is used to refer to something that you were obliged to do in order to comply with some authority, but in reality did not do.
3. *Should have* + past participle is used to refer to something that you did not do, but it would have been better if you had done it.
4. *Was going to* is used to refer to what you were planning to do but did not do.

	YES	NO
1	We **had to** perform six experiments to ensure repeatability.	We **musted** perform six experiments to ensure repeatability.
2	The manuscript **was supposed to have been completed** last week, but unfortunately they are still working on it.	The manuscript **had to be completed** last week, but unfortunately they are still working on it.
3	We **should have sent** the Abstract to the conference, then we could have presented our research. Now we can only go and watch.	We **had to send** the Abstract to the the conference, then we could have presented our research. Now we can only go and watch.
4	I **was going to send** my Abstract to the conference organizers, but I forgot.	I **had to send** my Abstract to the conference organizers, but I forgot.

12.8 obligation and recommendation: *should*

1. *Should (not)* is used to make strong recommendations (rather than giving direct orders).
2. *Should* is often found in Conclusions, when authors give their recommendations to other authors regarding possible directions for future work.
3. To avoid sounding arrogant, be careful how you use *should* when saying how your findings, applications, methodologies might be useful for other people – prefer *may.* Alternatively, precede your affirmation with *we believe that.*
4. In situations other than in papers, *should* is used to give friendly recommendations and to express opinions.
5. The form *ought to*, which has the same meaning as *should*, is rarely used in research. It often suggests a moral obligation.

	YES	AVOID
1	Special glasses **should** be worn in the lab. Computers **should not** be turned off without first being prepared for shut-down.	
2	Future work **should** address the need to …	Future work **must** address the need to …
3	Our approach **may** also be useful for those working in the field of medicine.	Those working in the field of medicine **should** also use our approach.
3	**We believe that** an important feature of any future work **should** be an attempt to …	An important feature of any future work **should** be an attempt to …
4	You **should** try using another search engine – it would be much quicker.	You **must** try using another search engine – it would be much quicker.
4	I think the third world debt **should** be cancelled.	I think the third world debt **has to** be cancelled.
5	There is a huge gap between what we feel we **ought to** do to help the third world, and what we actually do.	

13 Link words (adverbs and conjunctions): *also, although, but* etc.

13.1 *about, as far as ... is concerned*

1. Do not use *about* at the beginning of a sentence to introduce a topic.
2. As *far as x is concerned* is used to introduce a new topic in which the dependent phrase has a subject that is different from the topic introduced in the previous phrase.
3. Avoid unnecessary or excessive use of *as far as x is concerned*. It can often be rewritten in a more concise form.

	YES	NO (1), NOT ADVISED (2,3)
1	We are writing to you **about** the paper we sent you in May. We would like to	**About** the paper we sent you in May, we would like to know whether ...
	= Concerning / regarding / on the subject of / with regard to the paper we sent ...	
2	**As far as the budget is concerned**, **we** would to ask you whether ...	**As far as the budget is concerned**, **this** can be discussed at the next meeting.
	we is the subject of the second phrase	*budget* is the subject of both phrases
3	**The budget** can be discussed at the next meeting.	**As far as the budget is concerned**, **this** can be discussed at the next meeting.
3	**In terms of telephone production**, Nokia is Europe's biggest producer of mobile units.	**As far as telephones are concerned**, Nokia is Europe's biggest producer of mobile units.
	= Nokia is Europe's biggest producer of mobile telephones.	
3	**We can draw** a similar conclusion for the second phase as for the first phase.	**As far as the second phase is concerned** we can draw a similar conclusion as for the first phase.

A. Wallwork, *English for Academic Research: Grammar, Usage and Style*,
DOI 10.1007/978-1-4614-1593-0_13, © Springer Science+Business Media New York 2013

13.2 *also, in addition, as well, besides, moreover*

1. *In addition* is used to add an additional positive or neutral comment. *also, further, furthermore* can be used in the same way.

2. *Moreover* generally adds an additional negative comment – this is not a rule, but seems to be a preference among native English-speaking authors.

3. *Besides* and *in addition to* (both + *–ing* form) are used at the beginning of a sentence which is made up of two parts, in which the second part contains an additional feature or fact to the one given in the first part. *besides* is not used at the beginning of sentence to add an additional idea to the one presented in a previous sentence.

4. *As well (as)* means the same as *also*. *as well as* can be used at the beginning of a phrase and takes the *-ing* form of the verb. *as well*, but not *also*, can be used at the end of the phrase.

	YES	NOT RECOMMENDED (1,2), NO (3,4)
1,2	This software program has several interesting features .... **In addition / Also / Furthermore**, the cost is low and it is quick to learn.	This software program has several interesting features .... **Moreover**, the cost is low and it is quick to learn.
1,2	This software program has very few useful features. **Moreover**, the cost is very high and it is quick to learn.	This software program has very few useful features. **Further / In addition**, the cost is very high and it is quick to learn.
3	**Besides / In addition to** having several interesting features, this program is also economical ...	This software program has several interesting features .... **Besides**, the cost is low and it is quick to learn.
3,4	**In addition to / Besides / As well as** teach**ing** English, she **also** teaches French.	**In addition / Besides / As well to teach** English, she also teaches French.
4	She teaches French **as well**.	She teaches French **also**.
	She teaches French **as well as** English.	She teaches French **also** English.

13.3 *also, as well, too, both, all:* use with *not*

also, as well as, too, both and *all* are not generally found in negative sentences. So, in negative sentences:

1. Use *neither / nor* instead of *also, as well as, too.*
2. Use *either* instead of *both.*
3. Use *both* with a negative only for contrast.
4. Use *any* instead of *all.*

	YES	NO
1	X did **not** function and **nor / neither** did Y.	X did **not** function and **also** Y.
	Little is known about what truly matters in searching for information, **nor** what strategies users exploit.	Little is known about what truly matters in searching for information, **as well as** what strategies users exploit.
2	**Neither** of them **functioned** as required.	**Both** of them **did not function** as required.
3	We did **not** use **both** of them, just **one** of them.	We did **not** use **either** of them, just **one** of them.
4	There were no high scores in **any** of the tests.	There were no high scores in **all** the tests.

13.4 *although, even though* versus *even if*

1. *Even if* is only used for hypothetical situations, typically in second conditionals (9.2). Note: *also if* does not exist.
2. *Even though* and *although* have the same meaning. They are used to refer to real situations. They are generally found with present tenses. *though* means the same, but is not found at the beginning of a sentence in academic writing; *even though* is generally found at the beginning of a sentence rather than the middle.

	EVEN IF	EVEN THOUGH
1	**Even if** I was the President of the United States …	**Even though** researchers don't earn much money, at least they get to travel a lot.
2	**Even if** the book were available in English (it is currently only in Spanish), nobody would read it.	**Even though / Although** the book is essentially for children, adults still love to read it.

13.5 *and, along with*

1. In a list of three items or more, put a comma before *and* – this signals to the reader that the *and* is introducing the last item.
2. When giving a list of items, use semi colons (or commas) to highlight what elements *and* joins together.
3. When you use *and* several times within the same phrase, consider either rephrasing the sentence. Alternatively, use *along with* or *together with* to make your meaning clear.
4. *Along with* is followed by a noun. It can be used at the beginning of sentence to mean *in addition to* (13.2). *besides* (13.2) has the same meaning and can be used with a noun or verb.

	YES	NO
1	These countries include Tajikistan, Uzbekistan**, and** Kyrgyzstan.	These countries include Tajikistan, Uzbekistan **and** Kyrgyzstan.
2	The following groups of countries will be involved in the project: Tunisia and Egypt**;** Vietnam and Laos**;** Peru and Chile**;** and Poland and Estonia.	
3	I could visit your lab in **January**. I could also come in **February and March** if my professor agrees.	I could visit your lab in **January and February and March** if my professor agrees.
3	A and B, **along with** C and D, are the most used solutions.	A **and** B **and** C **and** D are the most used solutions.
4	**Along with / Besides** Spanish and Chinese, English is the most spoken language in the world.	**Along with** speaking English, she also speaks Hindi and Arabic.

13.6 *as* versus *as it*

1. When *as* is used without a following pronoun or noun, it has a similar meaning to *like* and *how.*
2. When *as* is followed with a pronoun (often *it*) or a noun it means *because* or *since.*

	AS	AS IT
1,2	This is not true, **as is** evident from the figure.	This is not true, **as / because it is** impossible to prove that X=Y.
1,2	**As mentioned** above and **as can** be seen in the figure …	These experiments were not performed **as / because it would** have required too much additional computing power.

13.7 *as* versus *like (unlike)*

1. *As* is used when the sense is that one thing is equal to another.
2. *Like* means 'similar to'.
3. *Unlike* is used when making a contrast. Note *differently from* does not exist in English, use *unlike* instead.

	AS	LIKE, UNLIKE
1,2	He works **as** a researcher in Paris.	She works **like** a slave for her boss.
	He is a researcher.	She is not a slave.
1,2	Diabetes acts **as** a significant risk factor for many physical diseases.	Xerostima: A symptom that **acts like** a disease.
	Diabetes is a risk factor.	Xerostima, i.e. dry mouth resulting from absent saliva flow, is not a disease but a symptom that can lead to a disease.
3,1	**As with copper and iron** techniques, lead substitution failed to demonstrate growth patterns in G. cirratum and C. altimus vertebrae.	Zinc, **unlike copper and iron**, fails to stimulate lipid peroxidation in vitro.
	Lead behaves in the same way as copper and iron.	Zinc does not behave in the same way as copper and iron.

13.8 *as, because, due to, for, insofar as, owing to, since, why*

1. *Because* indicates a consequence, *why* gives the reason or explanations.
2. *Because* can be used at the beginning of a sentence in order to explain a reason for doing something, but is usually replaced in formal English by *since, as, seeing as, given that, given the fact that, on account of the fact that* or *due to the fact that.* Another alternative is to use *in order to* or *so that.*
3. *Due to* and *owing to* mean the same as *because of*, and are followed by a noun. *owing to* tends only to be used at the beginning of a sentence.
4. *For* generally replaces *due to* and *because of* in phrases containing the word *'reason'.*
5. *Due to the fact that* and *owing to the fact* that are used before a subject + verb construction.
6. *Insofaras* and *inasmuchas* (also written *insofar as, in so far as, inasmuch as, in as much as*) can be used to replace because or due to the fact that when these appear at the beginning of a sentence. But they are somewhat antiquated.

	USAGE	ALTERNATIVE
1	This battery may explode when used with a third-party power supply. This is **because** the battery is highly inflammable and this is **why** it should not be used in children's toys.	This battery may explode when used with a third-party power supply. This is **due to the fact that** the battery is highly inflammable and this is **the reason [why]** it should not be used in children's toys.
2	**Because** they wanted total control, the revolutionary party enacted a series of drastic reforms.	**As / Since / Given that / On account of the fact that** they wanted total control, the revolutionary party enacted a series of drastic reforms.
		In order to have total control ...
		So that they would have total control ...
3	This accident was **due to** an electrical fault.	**Owing to** an electrical fault there was an accident.
4	The evolution of the Internet did not occur homogeneously around the world, **for** obvious historical, economic and political **reasons**. Moreover, **for reasons** of space we can only mention the ...	
5,6	**Due to the fact / Owing to the fact** we only had a limited budget, it was decided to use the cheapest version.	**Inasmuch as** we only had a limited budget, it was decided to use the cheapest version.

13.9 *both ... and, either ... or*

These expressions are frequently confused, thus leading to ambiguity for the reader.

1. *Both ... and* is inclusive.
2. *Either ... or* is exclusive. You cannot use *either* in both parts.
3. *Both* is only used with *not* when used to contrast.
4. *Not ... either ...* or indicates that none of the options are available.
5. The position of the preposition changes the meaning.

	YES	NO
1	We can go to **both** Iran **and** Jordon.	We can go **either** Iran **either** Jordon
	We will visit two places.	
2	We can go to **either** Iran **or** Jordon.	We can go **or** to Iran **or** Jordon.
	We can only visit one of the two alternatives.	
3	We ca**n't** go to **both** Iran and Jordon, but only to Iran.	
	We only have one choice.	
4	We ca**n't** go **either** to Iran **or** Jordon.	We ca**n't** go **neither** to Iran **nor** Jordon.
	We cannot visit these two places.	
5	We had fun **in both** the parks we visited and also the museums.	
	We visited two parks.	
	We had fun **both in** the parks and the museums.	
	We visited an undisclosed number of parks and museums.	

13.10 *e.g.* versus *for example*

1. Use a comma before *for example* and at the end of the example itself.
2. If you write *for example* after the example, rather than before, then it should be preceded and followed by commas. *for example* should not be placed at the end of the phrase.
3. In the middle of a sentence *e.g.* tends to be used for lists that are in brackets.
4. Don't use both *such as* and *for example* together. Use one or the other.
5. *For instance* and *like* are not normally used in research papers, prefer *for example.*

	YES	NO
1	Whenever you use your **PIN, for** example to get money from an **ATM,** do not let anyone see you.	Whenever you use your **PIN for** example to get money from an **ATM** do not let anyone see you.
2	Many governments are in crisis. In **Venezuela, for example, the** government is facing ...	Many governments are in crisis. In **Venezuela for example the** government is facing ...
		Many governments are in crisis. In Venezuela the government is facing big problems with the unions, **for example.**
3	When you use a PIN **(e.g.** to get money from an ATM, to pay for online purchases) ensure that ...	When you use a PIN **e.g**. to get money from an ATM, to pay for online purchases ensure that ...
4	We have collaborations with universities in many countries in Europe, **for example** France and Spain.	We have collaborations with universities in Europe, **such as for example** France and Spain.
5	We have given poster sessions at conferences in many countries in Europe, **for example** France and Spain.	We have given poster sessions at conferences in many countries in Europe, **like** France and Spain.

13.11 *e.g., i.e., etc.*

1. It is not necessary to put a comma immediately after *e.g.* and *i.e.*
2. *E.g.* and *i.e.* are also often written simply as *eg* and *ie*, but this may look confusing, particularly for non-native English readers.
3. *E.g.* introduces an example of what you have just said.
4. Use *i.e.* when what follows is a definition or clarification of what you have just said.
5. *E.g.* and *i.e.* are often confused. If you think your readers might not be familiar with the difference use *for example* and *that is to say*, respectively.
6. When you introduce a series of examples with *for example*, do not put *etc.* at the end.
7. If possible, think of something more meaningful than *etc.*
8. *Etc.* only requires one period (.) at the end of a sentence.

	YES	NO
1	Several authors, **e.g.** Schmidt, Si, and Hurria, have investigated this problem.	Several authors, **e.g.,** Schmidt, Si, and Hurria, have investigated this problem.
2	Several foods produce very strong allergies (**e.g.** eggs, nuts, wheat) …	Several foods produce very strong allergies (**eg** eggs, nuts, wheat **etc.**) …
3	This is true in at least ten countries, **e.g.** Spain, Japan and Togo.	This is true in at least ten countries, **i.e.** Spain, Japan and Togo.
4	The UK is made up of four countries, **i.e.** England, Scotland, Wales and N. Ireland.	The UK is made up of four countries, **e.g.** England, Scotland, Wales and N. Ireland.
5	The UK is made up of four countries, **that is to say** England, Scotland, Wales and N. Ireland.	
6	This is true in at least ten countries, **e.g.** Spain, Japan and Togo.	This is true in at least ten countries, **e.g.** Spain, Japan, Togo, **etc**.
7	This is true in many nations (Honduras **and other Central American countries**) and has very serious consequences.	This is true in many nations (Honduras **etc.**) and has very serious consequences.
8	This is true in at least ten European countries: France, Belgium, Sweden **etc**.	This is true in at least ten European countries: France, Belgium, Sweden **etc**..

13.12 *for this reason* versus *for this purpose, to this end*

1. *For this reason* explains why something was done.
2. *For this purpose* and *to this end* can be used indifferently to describe how something just mentioned was achieved.

	FOR THIS REASON	FOR THIS PURPOSE, TO THIS END
1,2	They wish to improve their English. **For this reason**, they are studying ten hours a day.	Our aim was to achieve higher performance. **For this purpose** we built an ad hoc device to provide increased power.
1,2	The patient was suffering from amnesia, **for this reason** it was difficult to question him directly on the circumstances of the accident.	It is now considered expedient to purge bone marrow of tumor cells prior to returning it to the patient, and **to this end** a variety of techniques have been developed.

13.13 *the former, the latter*

1. Only use *the former* and *the latter* when it is 100% clear to the reader what *the former* and *the latter* refer *to*. It is not bad style to repeat the key word, particularly as this will make it easier for the reader to identify exactly what is being referred to.
2. It may not be clear which element *the former* and *the latter* refer to. For example, when there are three elements, it may not be clear if *the latter* refers to the third element alone, or the second and the third.
3. In long sentences the reader may have already forgotten which elements were mentioned earlier.

	YES	NOT 100% CLEAR
1	**Lagos** and Khartoum are the capital cities of Nigeria and Sudan. **Lagos** has a population of …	**Lagos** and Khartoum are the capital cities of Nigeria and Sudan. **The former** has a population of …
2	In this recipe we used potatoes, carrots and beans. This is common practice with this kind of cooking. **The beans** can, of course, be steamed.	In this recipe we used potatoes, carrots and beans. This is common practice with this kind of cooking. **The latter** can, of course, be steamed.
3	Such an unsolicited bandwidth request can be **incremental** or **aggregate**. If it is **aggregate**, the X indicates the whole connection backlog. Blah blah blah blah blah blah blah blah blah blah blah blah blah blah blah blah blah blah blah. On other hand, if it is **incremental**, the X indicates the difference between its current backlog and the one carried by its last bandwidth request.	Such an unsolicited bandwidth request can be incremental or aggregate. In **the latter** case, the X indicates the whole connection backlog. Blah blah blah blah blah blah blah blah blah blah blah blah blah blah blah blah blah blah blah. In **the former** case, the X indicates the difference between its current backlog and the one carried by its last bandwidth request.

13.14 *however, although, but, yet, despite, nevertheless, nonetheless, notwithstanding*

1. To qualify what you have just written, use *however* (or *but*, which is slightly less formal). *however* is used in preference to *but* at the beginning of a sentence. *however* can be used with or without a comma, and can be located mid phrase between two commas. *nonetheless* and *nevertheless* are synonyms and mean the same as *however.*

2. *Yet* means the same as *but* and *however*, but has a stronger note of surprise. *still* has a similar meaning.

3. *Despite* and *notwithstanding* cannot be used when immediately followed by a *noun + verb* construction. Instead they have to be accompanied *by the fact that*. Thus, given that they are more complex to use, it is probably best to use *but, however* and *although* (13.4).

	YES	ALTERNATIVE	NO
1	The system costs very little to implement, **but / however / nevertheless / although** it is very complicated to use.	**However / Nevertheless**, it is very complicated to use. It is, **however / nevertheless**, very complicated to use.	The system costs very little to implement, **despite / notwithstanding** it is very complicated to use.
2	Governments know this is a problem, **yet** they do nothing about it.	**Although** governments know this is a problem, they **still** do nothing about it.	Governments know this is a problem, **despite** they do nothing about it.
3	**Despite** being cheap, the system works well.	**Although** the system is cheap, it works well.	**Although / Notwith-standing** being cheap, the system works well
3	**Despite the fact / Not-withstanding the fact** that the system is cheap, it is very effective.	**Although** the system is cheap, it is very effective.	**Despite / Notwithstanding** the system is cheap, it is very effective.
3	**Despite / Notwith-standing** the cheap price, the system works well.	The system works well **despite** its low cost.	**Despite** the cost is cheap, the system is very effective.
3	The system works well, **nevertheless** it is rather complicated.	The system works well, **however** it is rather complicated.	The system works well, **notwithstanding** it is rather complicated.

13.14 however, although, but, yet, despite, nevertheless, nonetheless, notwithstanding (cont.)

4. Only *however*, *nevertheless* and *nonetheless* can be used at the end of a phrase.

5. *However*, and *nevertheless* / *nonetheless* can be used at the beginning of a sentence, and be followed by a comma (13.15).

As highlighted by many of the examples, *although* (13.4) can often be used to qualify a statement. However, it is not used (a) between commas, (b) directly before a verb, (c) at the end of a phrase

	YES	ALTERNATIVE	NO
4	The system took only two days develop, it works well **nonetheless**.	The system was designed and … it works well **however**.	The system was designed and … it works well **despite / although**.
5	The system is cheap. **However**, it is difficult to implement.	The system is cheap. **Nevertheless / Nonetheless**, it is difficult to implement.	The system is cheap. **Notwithstanding / Despite**, it is difficult to implement.

13.15 *however* versus *nevertheless*

1. There is a very subtle difference between *however* and *nevertheless*. *however* can be used to add an additional observation or piece of information. *nevertheless* makes a stronger back-reference to what was said earlier, rather than focusing on giving new information. Essentially, if there is a causal relationship between two sentences, use *nevertheless*; otherwise use *however*.

	HOWEVER	NEVERTHELESS
1	Fewer men now seem to see career success as a central life interest around which other life activities are subordinated, **however** for many women the opposite is often true.	Studies indicate that stress from working long hours causes high blood pressure, **nevertheless / despite this** companies still insist on their employees working up to 60 hours per week.
	There is no direct correlation between the fact that fewer men are obsessed by their career and the fact that women now are – *nevertheless* cannot be used here.	There is a direct correlation between the fact that long hours are detrimental to health, but people continue to work 60 hours in any case. *however* could also be used here but the contrast would be weaker.
1	We didn't discuss your paper. **However** we did mention the possibility of you working in their lab.	We didn't discuss your paper. **Nevertheless**, there will be many other opportunities to talk about it.

13.16 *in contrast with* vs. *compared to, by comparison with*

1. Use *in contrast with* when the difference you are referring to is striking or surprising.
2. Use *compared to / with* and *by comparison* in all other cases.

	IN CONTRAST TO	COMPARED TO
1	**In contrast to** what was previously observed by Heimlich [2], our results showed an opposite trend.	**Compared to** Smith's results, our results are somewhat disappointing.
		= Our results are somewhat disappointing **by comparison with** Smith's.
2	**In contrast to** Hill's top-down approach [Hill, 2015], we start from the bottom layer.	**Compared to** the old technology, the new technology offers several new features.

13.17 *instead, on the other hand, whereas, on the contrary*

1. Use *instead* at the beginning of a sentence to resolve a problem stated in the previous sentence.
2. Do not use *instead* to introduce a new topic, even if the new topic is related in some way to the previous topic. Use *on the other hand.*
3. Use *on the other hand* to give an alternative or to add additional information about the thing mentioned previously – *whereas* is not used in such circumstances.
4. Both *on the other hand* and *whereas* can be used to make a contrast, but *whereas* gives the reader the idea that the contrast is quite strong. *whereas* is not normally used at the beginning of a sentence.
5. Do not use *on the other hand* simply to introduce new information without any sense of contrast.
6. *On the contrary* is only used to totally contradict what another author has stated.

	YES	NO
1/4	Do not join two independent clauses with a semicolon. **Instead**, make two simple separate sentences.	Do not join two independent clauses with a semicolon. **On the contrary**, make two simple separate sentences.
2	Italian and Spanish are similar languages, in fact they both derive from Latin. **On the other hand,** German is derived from …	Italian and Spanish are similar languages, in fact they both derive from Latin. **Instead,** German is derived from …
3	The conference may be held in Jordon, **on the other hand** it may be held in Egypt.	The conference may be held in Jordon, **whereas** it may be held in Egypt.
4	This year the conference is being held in Prague, **whereas** last year it was held on the other side of the globe in Sydney.	This year the conference is being held in Prague, **on the other hand** last year it was held on the other side of the globe in Sydney.
4	Italian and Spanish are similar languages, **whereas** German is completely different.	Italian and Spanish are similar languages. **Whereas** German is completely different.
4	We found that x = 1, **whereas [on the other hand]** Smith et al. reached a very different conclusion that x = 2.	
5	Much research has been carried out in the US on using sea animals as models for robots. **In addition / Furthermore**, new developments have been made in Japan with local species.	Much research has been carried out in the US on using sea animals as models for robots. **On the other hand**, new developments have been made in Japan with local species.
5	Italian and Spanish are similar languages, in fact they both derive from Latin. German, **on the other hand**, is derived from …	Italian and Spanish are similar languages, in fact they both derive from Latin. German, **instead / on the contrary**, is derived from …
6,4	Smith [2013] states that governments must intervene in such cases. We believe, **on the contrary**, that they absolutely must not intervene.	Smith [2013] states that governments must intervene in such cases. We believe, **whereas**, that they absolutely must not intervene.

13.18 *thus, therefore, hence, consequently, so, thereby*

1. *Thus, therefore, consequently, so* and *hence* all have the same meaning. They are used to indicate a consequence of what has just been said before. *so* is considered informal and is thus used less often.
2. *Hence* is generally reserved for mathematics.
3. *Thereby* means *in such a way*. It can only be used in the dependent phrase and is followed by a verb.

	YES	ALTERNATIVE
1	Researchers do not have much time to read papers. **Consequently**, it makes sense to write papers in a way that they can understand quickly and easily.	Researchers do not have much time to read papers. **Therefore / Thus**, it makes sense to write papers in a way that they can understand quickly and easily.
1	**Thus** the best way to write a paper is to use short sentences.	The best way to write a paper is **thus** to use short sentences.
2	Note that the right-hand side of equation (2) equals r(p)v(x) + [3]. **Hence**, equation (2) reduces to equation (1) if …	The square of the slope of the beam can be neglected in comparison with unity, **thus** equation (1) reduces to an ordinary linear equation.
3	Love promotes well-being **thereby** enabling people to live better lives.	Love promotes well-being **thus** enabling people to live better lives.

13.19 omission of words in sentences with *and, but, both* and *or*

You can omit certain words when used in conjunction with *and, but, both* and *or*. This helps to avoid unnecessary repetition

1. Nouns, pronouns, articles, possessive adjectives, *this, those* etc.
2. Verbs.
3. Prepositions.

	YES	YES
1	We measured and **[we]** calculated the values.	We extracted **[the fluid]** and then froze the fluid.
1	Give me your name and **[your]** address.	We need those books and **[those]** papers.
1	The sample can be introduced into the furnace using either a chromatographic **[pump]** or a peristaltic pump.	
	NB do <u>not</u> say: chromatographic **pump** or a peristaltic **one**.	
1,3	Is it a theoretical **[problem]** or **[a]** practical problem?	These can be found both in animals and **[in]** humans.
2	The flame was low but **[it was]** steady.	This is an expensive **[way]** but **[it is an]** effective way of reducing pollution.
3	This disease is predominantly found in the Sudan and **[in]** Chad.	These findings were true for adults and **[for]** children.

14 Adverbs and prepositions: *already, yet, at, in, of* etc.

14.1 *above (below), over (under)*

1. *Above* and *below* are typically used in a paper to refer to the location of sentences, paragraphs, figures and tables; *above* and *below* are also used when referring to levels, lists, averages and hierarchies.
2. *Over* has a similar meaning to *cover*, i.e. there is often physical contact between two elements.
3. *Over* and *under* also have a similar meaning to *more than* and *less than*, respectively.
4. *Under* also means 'in conformance with'.
5. Note the difference between *above all* (i.e. the most important thing) and *over all* (i.e. globally).

	ABOVE, BELOW	OVER, UNDER
1,2	As mentioned **above** there are three main methods, which are summarized in the table **below**:	A sheet was placed **over** the patient's body.
1,3	Pisa is 50 m **above** sea level which is **below** the national average for Italian cities.	Only children **over** the age of 13 were considered in the sample. Those **under** 12 years of age will be the subject of a future investigation.
4		**Under** the new regulations, all such documents have to be filed **under** 'funds'.
5	Many points need to be considered, **above all** age and sex.	**Overall**, our results can be considered as an important step towards finding a cure for this endemic disease.

A. Wallwork, *English for Academic Research: Grammar, Usage and Style*,
DOI 10.1007/978-1-4614-1593-0_14, © Springer Science+Business Media New York 2013

14.2 *across, through*

1. *Across* indicates the joining of two points on a plane.
2. *Through* indicates a transversal motion with some kind of penetration.
3. *Across* also means 'not restricted to one particular area'.
4. *Through* can also be used to mean *by means of.*

	ACROSS	THROUGH
1,2	They swam **across** the river.	The train went **through** the tunnel.
1,2	They walked **across** the road.	The sample was filtered **through** a very fine mesh.
3,4	Our method can be applied **across** disciplines.	We learnt this **through** lengthy research.

14.3 *already, still, yet*

1. *Already* at some time in the past. Do not confuse *already* with *just*. *just* means something that happened very recently (possibly a few seconds ago), e.g. *we have just arrived at the airport.*
2. *Yet* is frequently found in the affirmative and negative forms and refers to a period that started in the past and progresses up to (and possibly beyond) the present moment.
3. *Still* has the same meaning as *yet*, but is stronger. It indicates that a situation has not changed and may suggest surprise or concern.
4. *Already, yet* and *still* can also be used with the past perfect to put two past events in relation.

	ALREADY	YET	STILL
1,2,3	This procedure has **already** been explained elsewhere [Ying, 2013].	Has our paper been reviewed **yet**? Your paper has not been reviewed **yet**, and is scheduled for review on 2 June.	We **still** haven't heard from the referees. I am worried that they never received the paper, though I suppose they are **still** in time to contact us.
1,2,3	As **already** mentioned (see Sect 2.3), this method consists of …	As **yet**, no progress has been made in this field … = No progress has been made **yet**.	Despite sustained pressure by the democratic movement, his dictatorship **still** survives intact.
4	We **had already seen** her presentation before so we **did not want** to go again,	When we got the conference room, the presenter **had not arrived yet.**	Twenty minutes later, the presenter **had still not arrived.**

14.4 *among, between, from, of* (differentiation and selection)

1. *Between* when talking about a well-defined or well-separated number of items. It is found with verbs such as *decide, differentiate, discriminate, distinguish*, *mediate* and *synchronize*, and nouns such as *agreement, comparison, difference, distinction, interaction* and *relationship*. This is because such verbs and nouns indicate that a known number of items are involved.
2. *Among* when the group of items is not easily separable or the number is not known or is not important.
3. *From* with verbs such as *choose, pick, select,* and adjectives such as *different.*
4. *Of* is found at the beginning of a sentence when introducing a particular element that is part of a group.
5. *Of* is used when choosing from a number.

	AMONG	BETWEEN
1,2	The money was divided up **among** the participants.	The money was divided up **between** the three winners.
1,2	Students were encouraged to discuss their assignments **among** themselves.	We found no interaction in the classroom **between** teachers and students.
	= with each other	
1,2	Their paper discusses caste and social stratification **among** Muslims in India.	We analyse the relationships **between** Hindus and Muslims in India.
1,2	Many species have died out, **among** them X, Y and Z.	The two parties will have to sort out the differences **between** them.
	FROM	**OF**
3,4	Candidates will be chosen **from** diverse disciplines and then selected **from** a shortlist of 10.	**Of** the three candidates we interviewed, the last was certainly the best.
3,4	We selected our samples **from** a collection of 4543 items.	A comparison **of** the three figures reveals that …
5		Two thirds **of** tropical soils are oxisols and ultisols.
		Nine out **of** ten.
		Half **of** the samples.

14.5 *at, in, to* (location, state, change)

1. *At* before buildings and work place, *in* before towns, countries etc. In both cases no movement is involved.
2. *To* after a verb that indicates a destination.
3. *At* when describing the location of items in diagrams and figures, *in* before figures, tables etc. when used in association with verbs such as *see, show, highlight.*
4. *To* to indicate movement, change, conformance, limits and consequence: *adhere, adjust, attach, attract, bind, bring, come, confine, conform, connect, consign, convey, deliver, direct, email, fax, go, lead, link, move, react reply, respond, restrict, send, stick, supply, switch, take, tend, tie, transmit, write, yield.* This rule also applies to the related nouns: *delivery, modification, response, tendency* etc.
5. *In* is used before certain states e.g. *equilibrium, parallel, series.*
6. *To* is also used after certain adjectives that indicate position: *adjacent, close, contingent, contiguous, external, internal, next, orthogonal, parallel, perpendicular, tangent, transverse.*

	AT	IN	TO
1,2	They arrived **at** the airport, while we were still **at** work and Pete was **at** the restaurant.	They arrived **in** New York, while we were still **in** China.	They have gone **to** Beirut for a conference.
3,4	This can be seen **at** the top / bottom / side / edge of the figure.	As can be seen **in** Figure 1, the trend is … Also, as highlighted **in** Table 3 …	This was then moved **to** the top / bottom of the list.
5,6		The devices are placed **in** parallel and operate **in** a steady-state manner.	The lines are parallel **to** each other.

14.6 *at, in* and *on* (time)

1. *At* with a time of day, and with specific periods (*the weekend, Easter, Christmas*).
2. *In* with a period of time (week, month, year, decade, century etc.), including historical periods (*in the Middle Ages, in the Renaissance* etc.), and with *meantime / meanwhile.*
3. *On* with a day. Note some native speakers say *on the weekend* others *at the weekend.*

	AT	IN	ON
1,2,3	The meeting is scheduled to start **at** 15.30.	The conference will be held **in** June.	I will contact you **on** Monday morning.
1,2,3	We usually take our holidays **at** Easter or **at** Christmas, and of course **at** the weekend.	The last conference on this topic was held **in** 2012 and the previous one **in** the 1990s. The first was held **in** the 18th century.	We do not work **on** Christmas Day, **on** Easter day and **on** July 4 (Independence Day).

14.7 *at, to* (measurement, quality)

1. *At* with the following nouns that indicate quantity and measurement, e.g. *degree, interval, level, node, point, pressure, ration, speed, stage, temperature, velocity.*
2. *To* with the other types of calculations and measurements, e.g. with the following verbs: *approximate, calculate, correct, heat, measure, raise.*
3. With certain adjectives *to* indicates a quality, conformance or similarity: *inferior, superior; equal, identical, proportional, similar; immune, impermeable, open, resistant, sensitive*; *according, alternative, analogous, attention, common, comparable, conformance, compliance, correspondence, entitlement, identical, inferior, likened, open, opposed, proportional, relative, relevant, responsive, similar, suited, superior, transparent.*

	AT	TO
1,2	Water boils **at** a temperature of 100 C.	Heat the water **to** a temperature of 50 C.
1,2	The vehicle moves **at** a velocity of 300 cm / h.	The potassium content was approximated **to** 90 mEq / kg.
3		Gender is common **to** all Latinate languages, but has no adherence **to** logical rules.

14.8 *before, after, beforehand, afterwards, first* (time sequences)

1. *Before* and *after* must precede either a noun / pronoun, a gerund, or an entire subordinate phrase.
2. *Before* and *after* cannot be used as conjunctions or adverbs. Instead use *beforehand* and *afterwards.*
3. *First* means 'before anything else', it is often followed by *second(ly)* or *then*. It is thus used to list a sequence of actions.

	BEFORE / AFTER	BEFOREHAND / AFTERWARDS	FIRST
1,2,3	Where are you going **after** the congress?	We're going for a drink and **afterwards** back to the hotel.	**First** we are going for a drink, **then** afterwards back to the hotel.
1,2,3	**Before** checking the levels, the presence of any metals should be detected.	The solution consists in detecting the presence of metals **beforehand** and then / subsequently checking the levels.	**First(ly)** we detect the metals, **secondly** we check the metals, and finally we …
1,2,3	Preparations should be made **before** the mixture becomes solid.	Preparations should be made **beforehand**. = made in advance	**First** the preparations should be made, **then** the mixture should be allowed to become solid.

14.9 *beside, next to, near (to), close to* (location)

1. *Beside* and *next to* have the same meaning and indicate elements that are touching or almost touching.
2. *Near (to)* and *close to* have the same meaning and indicate elements that are at some distance to each other.
3. *Nearby* / *close by* replace *near* and *close* at the end of a phrase.

	BESIDE, NEXT TO	NEAR TO, CLOSE TO
1,2	I sat **beside / next to** her at the conference.	Our hotels were quite **near to / close to** each other, but on opposite sides of the river.
3		There was a train station **nearby / close by.**

14.10 *by* and *from* (cause, means and origin)

1. *By* when the agent of an action is mentioned.
2. *From* when the origin is mentioned. Verbs typically followed by *from* are: *arise, benefit, borrow, deduce, defend, deviate, differ, ensue, exclude, originate, profit, protect, release, remove, select, separate, shield, subtract, suffer.* Likewise, some of the nouns that derive from these verbs are also found with *from: deviation, exclusion, protection.*
3. *By* when the method or means is given.
4. *From* is often found with *to*, to indicate the move from one place to another.

	BY	FROM
1,2	Our paper has now been revised **by** a native English speaker.	We received a letter of acceptance **from** the editor.
1,2	The original computers were made **by** IBM but were then replaced **by** the director.	This mixture is made **from** a variety of substances **from** all over the world.
1,2	Taxes were raised **by** the government.	The economic crisis arose **from** banking malpractices and indiscriminate consumer-borrowing **from** banks.
1,2	Considerable damage was caused **by** the earthquake.	This paper suffers **from** a lack of detailed discussion and would also benefit **from** a complete revision of the English.
1,2	The number was then divided or multiplied **by** 32.5, depending on the case.	The corresponding amount was obtained by subtracting the first value **from** the second.
3,2	They learned English **by** watching videos on YouTube.	They quickly learned English **from** their native-speaking colleagues.
3,4	We went **by** train instead of going **by** car or **by** plane.	While on the train **from** Malmo **to** Stockholm, they kept switching **from** one language **to** another.

14.11 *by, in, of* (variations)

When talking about increases, decreases, modifications, changes, variations etc., use:

1. *By* after a verb.
2. *In* with a noun.
3. *Of* with a number.

See also *with* (14.14.3)

	BY	IN	OF
1,2,3	The stock market has risen **by** 213 points.	There has been an increase **in** inflation.	There has been an increase in inflation **of** 5%.
1,2,3	Attendance has fallen **by** 16%.	A fall **in** unemployment is predicted.	This was affected by variations **of** 16% and more.

14.12 *by* and *within* (time)

1. *By* with an end date.
2. *Within* for a period.

	BY	WITHIN
1,2	We must receive your manuscript by January 21 or at the latest **by** the end of the month.	Manuscripts will be reviewed **within** six weeks of receipt.

14.13 *by now, for now, for the moment, until now, so far*

1. *By now* means 'given everything that has happened before'. *for now* and *for the moment* both mean 'from this point in time until some time in the future when a change is expected'. *for now* and *for the moment* are generally followed either by the **present simple** or ***will.***
2. *Until now* and *so far* both mean 'from a certain point in the past up to the present moment and possibly in the near future too'. Both are usually used with the **present perfect.**
3. *Until now* is not normally used directly before a past participle. Note *till*, which means the same as *until*, is considered too informal for research manuscripts.

	BY NOW	FOR NOW
1	It should, **by now**, be well known that publications in peer-reviewed journals are more likely to ensure success than …	We wanted to buy new equipment, but we do not have the funds, so **for now / for the moment** we will have to continue using our old equipment.
1	The literature on this topic should, **by now**, be extremely familiar to …	We shall expand more fully on this in Sect. 3. **For now**, we just focus on …
	UNTIL NOW	**SO FAR**
2	**So far / Until now**, research into this area has been limited to X. In this paper, we investigate Y.	This is the only acid that has **so far / until now** been found to be effective in such scenarios.
3		The research **so far** undertaken has only focused on …
		The patients **so far** described all had benign non-calcified nodules. Let us now turn to cases with calcified nodules.

14.14 *during, over* and *throughout* (time)

1. *During* means at some point in the course of a period of time. This period can either be in the past or future.

2. *Over* often refers to a period of time that began in the past and is still true in the present, *over* is thus normally used with the present perfect (8.3). However, *over* can also be used for a future period.

3. *Throughout* means for the entire course of a period of time. This period can refer to the past, present or future.

	DURING	OVER	THROUGHOUT
1,2,3	I hope to have the opportunity of meeting you **during** the conference next month.	**Over** the last few years, there has been increasing interest in … = For the last few years, i.e. until and including today	**Throughout** history, humans have had a tendency to collect objects – even objects of no apparent value.
1,2,3	I worked with him **during** my Erasmus project.	**Over** the last decade, no progress has been made in … However, **over** the next few years this will certainly change.	Plagues were common **throughout** the Middle Ages.

14.15 *for, since, from* (time)

The adverb of time you use will normally help you to understand the correct tense to use (Section 8).

1. *For* indicates the duration from the past until the present. It is typically used with plural words indicating time, e.g. *days, months, years, decades*. *for* answers the question 'how long has this situation been ongoing?' In this sense, *for* is used with the **present perfect**. Similar expressions denoting a duration from past to present are: *over* (e.g. *over the last two decades*), *so far, until now* (8.17).
2. If *for* is used to indicate a period of time that is now finished, then it is used with the **simple past.**
3. *Since* indicates the starting point of a current situation. It is typically used with precise points in time, e.g. *2001, last month, yesterday*. *since* answers the question 'when did this situation begin?'
4. *From* indicates a range of time, i.e. with a start and finish. Because there is a finish time, *from* is not used with the **present perfect**. But *from* can be used with most other tenses.

	YES	NO
1	We **have been doing** this research **for** nine years.	We **do** this research **from / since** nine years.
1	**Over the last few months** there **has been** a lot of media coverage.	Over the last few months there **is** a lot of media coverage.
2	I **studied** in Boston **for** three years and **then** I moved to Beijing.	I **have studied** in Boston **for** three years and **then** I **have moved** to Beijing.
3	**Since** 2001 there **has been** a dramatic increase in suicides.	**From / Since 2001** there **is** a dramatic increase in suicides.
4	I **studied** in Boston **from 2008 to 2011**.	I **have studied** in Boston **from 2008 to 2011**.

14.16 *in, now, currently, at the moment*

1. *In* is generally followed by a date (e.g. *in June, in 2016*) and is therefore not used with the **present perfect**. *in* can be used with the **present**, **future** and the **past** (e.g. *the new term starts in October, I will see you in March, I last saw her in 2011*).

2. *Now, currently, at the moment* indicate a time period that is ongoing, so they are used with the **present simple**. On the other hand, *ago, yesterday, last week* (*month, year*) indicate a time that is completely past so they must be used with the **simple past.**

3. Adverbs that indicate a connection between the past and present are generally found with the **present perfect** (*historically, traditionally, typically*) but *traditionally* and *typically* can also be used with the **simple present** and **simple past**, depending on the context.

	YES	NO
1	I **joined** this research group **in July**.	I **have joined** this group **in July**.
2	This **is currently** the world's biggest problem.	**Until last year this has been** the world's biggest problem.
3	**Historically**, French **has always been taught** in English schools as the second language.	**Historically**, French **was always taught** in English schools as the second language.

14.17 *in, inside, within* (location)

1. *In* and *inside* often have a similar meaning in relation to a confined space.
2. *Inside* is the opposite of *outside*, *within* is never used in this sense.
3. *Within* means internal to something that can be a real physical space (e.g. *border*), or abstract (e.g. *confines, framework, comprehension*).
4. *In* is not usually found at the end of the sentence after a verb.
5. *Inside* has a metaphorical meaning of 'revelations' regarding something, and is often found in paper titles (as in the examples given below).

<table>
<tr><th></th><th>IN</th><th>INSIDE</th><th>WITHIN</th></tr>
<tr><td>1</td><td>The money is kept in the safe.</td><td>The hostage was kept inside the same room for more than three years.</td><td></td></tr>
<tr><td>1,2,3</td><td>We examined the links among the parents of children in the school.</td><td>The children were only allowed to play inside the school, never outside.</td><td>Insufficient attention has been given to the importance of relationships among children within the school.</td></tr>
<tr><td>4,3</td><td></td><td>The children could never go outside, they were always kept inside.</td><td>The paper draws on research in six EU member states carried out within the framework of a project on climate change policies.</td></tr>
<tr><td rowspan="3">5,3</td><td rowspan="3"></td><td>Inside the mind of a monkey</td><td rowspan="3">Our aim is to present, within the limits of national security, a comprehensive summary of this data regarding the war in Iraq.</td></tr>
<tr><td>Inside bureaucracy</td></tr>
<tr><td>Inside the family</td></tr>
</table>

14.18 *of* and *with* (material, method, agreement)

1. *Of* indicates a material out of which something is made.
2. *With* explains how something is created or what something is equipped with.
3. *With* indicates the presence or absence of a relationship, agreement or support. *with* is used with the following verbs, nouns and adjectives: *accordance, acquaint, agree, ally, appointment, associate, coincide, collaborate, comparable**, compliance, comply, concur, connect**, connection**, consistent, contact, contaminate, cooperate, cooperation, coordinate, coordination, coupled, cover, deal, dispense, endow, entrust, equip, experiment, help, incompatible, incongruous, infect, interact, interfere, liaise, mix, paint, problem, provide, reinforce, synchronize, synchronous, tally* [* indicates that these words can also be followed by *to*].
4. *With* also means 'as a function of', *of* cannot be used in such cases.

	OF	WITH
1,2	The royal family were wearing jewels made **of** gold and silver.	A cake can be made **with** various ingredients.
1,2	Nitonol is an alloy **of** nickel and titanium.	These cars are manufactured **with** armor plating and come equipped **with** bullet-proof windows.
1,3	Snow is made **of** small crystals of ice.	The terrain was covered **with** snow.
4		The severity of the illness varies **with** age.

15 Sentence length, conciseness, clarity and ambiguity

15.1 maximum two ideas per sentence

1. Ideally, each sentence should contain only one piece of information and should be no more than about 25 words long.
2. The occasional long sentence is fine, provided that: (1) it is easy to understand, (2) dividing it into shorter sentences would be difficult to achieve.

Note: The 'No' examples were written by a native English speaker 200 years ago – this style of writing is no longer considered acceptable.

	YES	NO
1	The majority of words recorded in a modern English dictionary have been borrowed from other languages. However, the words ordinarily used in speaking are largely of English origin. Most words have somewhat changed in form since their first introduction into England. *(16, 13, 13 words)*	In the language as recorded in a modern English dictionary the great majority of words are borrowed; but the words we ordinarily use in speaking are largely of English origin, although for the most part somewhat changed in form since their first introduction into England. *(45 words)*
2	As has been shown above, it would be incorrect to say that English was derived from Latin, or French, or Greek, of from anything else but the original language of the Teutonic branch of the Indo-European language. Nevertheless Latin, French and Greek have had a great and lasting influence on English vocabulary. *(37, 15 words)*	Although (as has been shown above) it would be incorrect to say that English was derived from Latin, or French, or Greek, of from anything else but the original language of the Teutonic branch of the Indo-European language, nevertheless Latin, French and Greek have not been without great and lasting influence on our vocabulary. *(54 words)*

DOI 10.1007/978-1-4614-1593-0_15,

15.2 put information in chronological order, particularly in the methods section

Your document should be like a map showing the reader the direction to follow. Try to write in a step-by-step way, with each step logically following the previous one. This generally entails putting information in chronological order.

1. In the second part of a sentence prefer *then* + simple past, rather than *after* + past perfect.
2. Use *first(ly), second(ly)* etc. to show the stages of a procedure.

	YES	NO
1	The vegetables were **cooked** in the oven and **then served** with the main course.	The vegetables were **served** with the main course **after they had been cooked** in the oven.
2	When defusing a bomb, **first disrupt** the circuit. **Secondly**, cut the green wire.	When defusing a bomb, cut the green wire **after first having disrupted** the circuit.

15.3 avoid parenthetical phrases

Subjects often get separated from their verbs by parenthetical information contained between two commas or in brackets. In such cases, the use of commas and brackets breaks the flow of the sentence and makes it harder to understand immediately.

1. Rearrange the sentence so that the subject and verb are next to each other. The order you use will depend on the emphasis you want to give.
2. When the parenthetical information is rather long, split the sentence up.

	YES	AVOID
1	**This feature will** only be of limited use, owing to its high cost.	**This feature**, owing to its high cost, **will** only be of limited use.
	Owing to its high cost, **this feature will** only be of limited use.	**This feature** (owing to its high cost) **will** only be of limited use.
1	The **vegetables were cooked** in the oven **and then** served with the main course.	The **vegetables**, cooked in the oven, **were served** with the main course.
		The **vegetables,** which had been cooked in the oven, **were served** with the main course.
2	**We believe** the results are significant given their innovative nature. **When** they are analysed they should help in our understanding of the diffusion of this virus in the world today.	The analysis of the results, **which we believe are of a significant value given their innovative nature**, should help in the understanding of the diffusion of this virus in the world today.

15.4 avoid redundancy

Make it easy for the reader by using the minimum number of words. The resulting sentences should also be quicker for you to write.

Don't use:

1. Meaningless abstract words.
2. Meaningless descriptive words.
3. Unnecessary introductory phrases (13.1.3).
4. Unnecessary link words (13) e.g. *in particular, furthermore, to be precise.*

	YES	AVOID
1	This supports the **installation** of the software.	This supports the **activity of installation** of the software.
1	Achieving this is **difficult**.	Achieving this is **a difficult task**.
1	We believe the results are **significant**.	We believe the results are **of significant value.**
2	They should be **green and round**.	They should be **green in color and round in shape**.
3	**Note** that the sum of the values needs to be lower than …	**It is worth noting / Bear in mind** that the sum of the values …
4	We found that x = y. Under certain circumstances x also equals z.	We found that x = y. **In particular**, under certain circumstances x also equals z.

15.5 prefer verbs to nouns

1. Use a *verb* rather than a *noun*, this improves readability and conciseness.
2. Use one *verb* rather than a *noun* + *verb.*

	YES	AVOID
1	This was used to **calculate** the values.	This was used **in the calculation of** the values.
1	By correctly **choosing** the parameters, performance can be improved.	Through the correct **choice** of the parameters, performance can be improved.
1,2	This allows us **to transfer** the money.	This allows **the transfer** of the money to be **performed**.
	This allows the money **to be transferred**.	
2	The USA **was compared** to the Russian Federation.	A **comparison was made** between the USA and the Russian Federation.
2	The Russian Federation **performed** much better than the USA.	The Russian Federation **showed a** much better **performance** than the USA.

15.6 use adjectives rather than nouns

1. Use a *verb* + *adjective* construction rather than *verb* + *noun.*
2. Use *more* + *adjective*, rather than a *comparative adjective* + *noun.*

	YES	AVOID
1	This method has quite an **efficient** calculation process.	This method shows quite a good **efficiency** in the calculation process.
	= Calculations with this method are quite **efficient**.	
2	X is **more homogeneous** than Y.	X has **a higher homogeneity** with respect to Y.

15.7 be careful of use of personal pronouns: *you, one, he, she, they*

1. The use of *you* to address the reader directly is rare in research manuscripts. It is normally reserved for user guides, instruction manuals, websites and email.
2. The generic pronoun *one* is somewhat archaic and in any case can easily be avoided.
3. Only use *he* and *she* (and *his, him, her, hers*) when they are specifically used in relation to a male or female subject, respectively. Using either *he* or *she* to refer to a generic person (i.e. where the sex of the person is irrelevant) sounds either politically incorrect or simply strange.
4. You can avoid using *he* and *she* (and *he / she* etc.) by making the subject plural and using *they / their / them / theirs.*
5. When the subject must be singular, do not use the masculine pronoun, instead use *he / she, him / her* and *his / her.*

	YES	AVOID
1	A more detailed explanation **can be found** in Appendix B.	**You can find** a more detailed explanation in Appendix B.
2	This feature would be useful in many cases.	**One** can think of many examples where this feature would be useful.
3	Barack Obama claimed in **his** speech that **he** … whereas in **her** speech Angela Merkel reaffirmed that **she** …	A **doctor** plays a vital role in society, in fact **he** often …
		A **primary teacher** can have a great influence on the future lives of **her** pupils.
4	If **traders** are trading on several markets and **they** wish to …	If **a trader** is trading on several markets and **he** wishes to …
4	When **users** have connection problems, the system tries to reconnect **them** automatically.	When **a user** has connection problems, the system tries to reconnect **him / her** automatically.
5	There are two traders: Trader A and Trader B. If Trader A wants to send **his / her** order to the market then **he / she** has to …	There are two traders: Trader A and Trader B. If Trader A wants to send **his** order to the market then **he** has to …

15.8 essential and non-essential use of: *we, us, our*

The use of *we* may be essential to avoid ambiguity (10.4).

However, you can avoid using *we* (*us*, *our, ours*) in the following cases:

1. When you can be more concise by not using *we.*
2. When you are talking about people in general.
3. When you are talking about a procedure that you did not invent yourself. Instead, use the passive.
4. Sometimes authors use *we* to involve the reader in the logical process being described. However, if you find you are using *we* in almost every sentence consider using alternative solutions, for example, the passive or gerund.
5. A journal's ruling on not to use *we* may result in awkward phrases. In such cases *we* should, in my opinion, be used.

	YES	AVOID
1	This document **outlines** the main points of xyz.	In this document **we outline** the main points of xyz.
1	This means that **there are** two ways to solve this problem.	This means that **we have** two ways to solve this problem.
2	The last few years **have witnessed** a considerable increase in the numbers of mobile devices.	In the last few years **we have witnessed** a considerable increase in the numbers of mobile devices.
3	A cloze procedure is a technique in which **words are deleted** from a passage according to a word-count formula. The passage **is presented** to students, who insert words as they read. This procedure **can be used** as a diagnostic reading assessment technique.	A cloze procedure is a technique in which **we delete** words from a passage according to a word-count formula. **We present** the passage to students, who insert words as they read. **We can use** this procedure as a diagnostic reading assessment technique.
4	Before **dealing** with this issue …	Before **we deal** with this issue …
4	As already pointed out by Ying, this is valid only below a certain frequency (**hereafter** F).	As already pointed out by Ying, this is valid only below a certain frequency which we **denote by** F.
4	The discussion of integers **is now continued** to extend the notion of …	**We continue our** discussion of integers to extend the notion of …
5	**We believe** that this approach is both the easiest and quickest to perform.	**It is the authors' subjective impression** that this approach is both the easiest and quickest to perform.

15.9 avoid informal words and contractions

1. Some words and expressions relating to quantities are considered too informal for research papers.
2. Prefer *such as* to *like* when giving examples.
3. Prefer *thus* (*therefore, consequently*) to *so;* prefer *in any case* to *anyway.*
4. Avoid *actually* at the beginning of a sentence.
5. Prefer *until* to *till.*
6. Avoid contracted forms (e.g. *it isn't, we'll*). Instead use the full form (e.g. *it is not, we will*).
7. Avoid colloquial phrasal verbs (e.g. *check out, get around, give up, work out*). See 16.9.

	YES	GENERALLY CONSIDERED TOO INFORMAL
1	The sample size was **quite** small. A **few** samples were contaminated.	The sample size was **pretty** small. A **tiny part of the** samples were contaminated.
2	A few European countries, **such as** Montenegro, Slovenia and Moldavia, have requested …	A few European countries, **like** Montenegro, Slovenia and Moldavia, have requested …
3	The first set of samples were contaminated. We **thus** had to … **In any case**, this was useful because …	The first set of samples were contaminated, **so** we had to … **Anyway**, this was useful because …
4	His behavior was strange. **In fact**, he rarely talked …	His behavior was strange. **Actually**, he rarely talked …
5	We waited **until** the end of the experiments before …	We waited **till** the end of the experiments before …
6	**Let us** now turn to … One **cannot** but notice that … **We have** seen that …	**Let's** now turn to … One **can't** but notice that … **We've** seen that …
7	Clinton's argument does not appear to **make sense**, although Smith et al. have **defended** Clinton's position.	Clinton's argument does not appear to **add up**, although Smith et al. have **stuck up for** Clinton's position.
7	They tried for 20 years to prove that x=y and they finally **succeeded** in 2012 when some missing data were **discovered** by chance.	They tried for 20 years to prove that x=y and the finally **brought it off** in 2012 when some missing data **turned up** by chance.

15.10 emphatic *do* / *does*, giving emphasis with auxiliary verbs

1. Do and *does* are very occasionally used in an affirmative sentence to give emphasis.
2. Do not use *do* and *does* when no special emphasis is required.
3. If you want to emphasize other auxiliary verbs you can put them in italics or underline them and / or associate them with another expression giving emphasis (e.g. *however, instead, in fact*) – such usage is rare.
4. Do and *does* (and other auxiliary verbs) are also used in affirmative sentences with *only* and negative adverbs. In such cases subject and verb are inverted (16.6).

	EMPHASIS	NO EMPHASIS
1,2	Are scientists whose native language is not English at a disadvantage in attempting to get work published and accepted? Certainly, **there does seem** to be evidence that scientists from developing countries **do find** it more difficult to get their work published than those in developed countries.	In comparing the interviewees' responses with the comparable data from previous studies **there seems** to be evidence that this sample may have been affected by issues regarding the way the questions were posed. In fact, **we found** that three questions were open to several interpretations.
1,2	Such strange phenomena have been reported. Whether **they do in fact** indicate the existence of UFOs is still an open issue.	These drugs may not always be effective. **In fact, they often have** undesired side effects.
3	Contrary to what was previously thought, it ***is*** possible to automatically acquire English with a brain implant, and such an operation ***can***, in fact, be achieved at low cost.	
4	In most countries using plastic bottles is not considered a problem. **Only** in Scandinavia **do they** insist on using glass bottles.	In most countries using plastic bottles is not considered a problem. **However,** in Scandinavia **they** insist on using glass bottles.

15.11 ensuring consistency throughout a manuscript

1. Use the same format for the same word.
2. Use the same grammatical form.
3. Use the same spelling (27.2).

	YES	NO
1	In **Phase 1** of the project we will …, whereas in **Phase 4** we will …	In the **first phase** of the project we will …, whereas in **Phase IV** we will …
2	This research has three main aims: (1) **to increase** efficiency (2) **to enhance** existing features (3) **to lower** costs.	This research has three main aims: (1) **to increase** efficiency (2) **the enhancement** of existing features (3) **lowering** costs.
3	The **behavior** of the children in the **realization** of the truth differed radically from the **behavior** of their respective parents who **realized** the truth considerably more rapidly.	The **behavior** of the children in the **realization** of the truth differed radically from the **behaviour** of their respective parents who **realised** the truth considerably more rapidly.

15.12 translating concepts that only exist in your country / language

Avoid literal translations of concepts that are peculiar to your own language or country (political entities and traditions, names of festivities etc). Instead:

1. Use your language (in italics) and then explain in English.
2. Explain in English without using your language.

	YOUR LANGUAGE + ENGLISH	JUST ENGLISH	NO
1	Such preferences are due to the Chinese concept of ***yì tóu* which is related to ideas regarding premonition and superstition**. Thus …		Such preferences are due to the Chinese concept of ***yì tóu***. Thus … This is OK if you have already given an English explanation of *yì tóu or if you know your readers will be familiar with the term.*
1	Kimura [2014] regards ***aida* (literally 'betweeness')** as a transpersonal source that …		Kimura [2014] regards ***aida*** as a transpersonal source that …
	= Kimura [2014] regards ***aida* or 'betweeness'** as a transpersonal source that …		This is OK for the same reasons as explained in the previous example.
2	The Dutch celebrate **Sinterklaas (i.e. Santa Claus)** on December 5. Sinterklaas arrives simultaneously at every city or village in the Netherlands. This is explained by way of the so-called **"hulp-Sinterklazen"** (people who help Sinterklaas by dressing up like him).	… Sinterklaas manages to arrive simultaneously at every city or village with the aid of **helpers who dress up like him**.	The Dutch celebrate **Father Christmas** on December 5. He arrives simultaneously at every city or village in the Netherlands. This is explained by way of the so-called "**help Father Christmas**". Father Christmas is not 'celebrated'. 'Help father Christmas' is meaningless in English.
2	The energy plan was approved by the **Regione Toscana (i.e. the regional administration in Tuscany)** and was then …	The energy plan was approved by the **regional administration in Tuscany** and was then …	The energy plan was approved by the **Tuscany Region** and was then … 'Tuscany Region' is meaningless in English.

15.13 always use the same key words: repetition of words is not a problem

1. Never invent synonyms for key words. The reader may think that each term has a different meaning.
2. When you refer back to a key word used earlier and there are many words between the key word and what you are writing now, it is better to repeat that key word rather than using a generic form (e.g. *hydrogen … hydrogen*, rather than *hydrogen … this gas*). If you just use the generic term (e.g. *this gas*) or a pronoun (*it*), the reader may be forced to re-read the paragraph to understand what subject you are referring to.
3. It is not considered bad style to repeat the same preposition. Prepositions are not generally interchangeable and therefore only the correct preposition should be used.
4. It is perfectly acceptable to use synonyms for non-key words such as verbs (e.g. *carry out, perform*), adjectives (e.g. *important, crucial*) and adverbs (e.g. *often, frequently*).

	YES	NO
1	In the **first phase** of the project we will …, whereas in the **fifth phase** we will …	In the **first phase** of the project we will …, whereas in the **fifth stage** we will …
1	The **operator** of the PC does x followed by y. Finally, the **operator** does z.	The **operator** of the PC does x followed by y. Finally, the **user** does z.
2	The solubility and mobility of elemental **mercury** is low and blah blah blah, blah blah blah blah blah blah blah. In fact, blah blah blah blah blah. In addition, blah blah blah blah blah blah blah blah blah. However, **mercury** can undergo many transformations, leading to contamination in humans.	The solubility and mobility of elemental **mercury** is low and blah blah blah, blah blah blah blah blah blah blah. In fact, blah blah blah blah blah. In addition, blah blah blah blah blah blah blah blah blah. However, this **metal** can undergo many transformations, leading to contamination in humans.
3	There is unquestionably a need **for** methods **for** testing **for** synergy with combinations **of** any number **of** agents **of** a certain value.	
4	Such manuscripts are **normally** accepted, **usually** within 20 days of receipt.	

15.14 avoid ambiguity when using *the former / the latter, which*, and pronouns

1. Avoid using *the former* (13.12) and *the latter*, or *the first* and *the second*. Just repeat the words. This saves the reader having to re-read the paragraph to find out what *the former* or *latter* refers to.
2. Avoid relative clauses (7) if the resulting sentence would be too long. Instead repeat the key word.
3. When using pronouns (*it, they, them, one* etc.), make sure it is clear what these pronouns refer to.

	YES	NO
1	Examples of countries where this kind of election system is used are **Australia, New Zealand** and the **Canada**. From an analysis of the literature it would seem that **Australia** has been the target of most investigations into …	Examples of countries where this kind of election system is used are **Australia, New Zealand** and the **Canada**. From an analysis of the literature it would seem that the **former country** has been the target of most investigations into …
2	The CNR is the Italian National Research Council and has many institutes where innovative research is **carried out. These institutes** are located in various parts of Italy such as Pisa, Turin and Rome.	The CNR is the Italian National Research Council and has many institutes where innovative research is carried out and **which** are located in various parts of Italy such as Pisa, Turin and Rome.
3	Portuguese and Spanish are spoken more widely than French and Dutch. In fact, **French and Dutch** are only used in ex-colonies, although **French** is also spoken in …	Portuguese and Spanish are spoken more widely than French and Dutch. In fact, **they** are only used in ex-colonies, although **the first** is also spoken in …

15.15 avoid ambiguity when using *as, in accordance with, according to*

1. *As* is sometimes used to indicate whether an approach, method or practice conforms to the specifications (often legal) of such an approach, method or practice.

2. Ensure that it is clear whether you are saying that something conforms or does not conform to regulations and recommendations.

3. *In accordance with, according to, in compliance with* and other expressions indicating conformity to rules and regulations are subject to the same possible ambiguity outlined in Rule 1.

	YES	NO
1	**As recommended** by ISO 2564.89, we used the following procedure:	
1	We adopted the guidelines regarding land use planning, **as required** by Council Directive 96 / 82 / EC.	
2	In 16.6% of the food packages, **no** country of origin was reported. **This is in direct contrast** to European regulations, which explicitly state that the country of origin must be declared.	In 16.6% of the food packages, no country of origin was reported, **as required** by European regulations. It seems that stating the country of origin is not a European requirement.
3	**As suggested by / In accordance with** Gomez (2015), samples were not pre-washed.	The samples were not pre-washed **in accordance with** Gomez (2015).
	Contrary to what suggested by Gomez (2015), samples were not pre-washed. This decision was made because …	It is not clear whether Gomez advocates pre-washing or not.

15.16 when expressing a negative concept using a negation

1. English tends to express negative ideas with a negation (17.7). This helps the reader to understand immediately that something negative is being said. Note that the examples given in the right-hand column would be perfectly acceptable in a manuscript, those on the left would be better in an oral presentation.
2. Only one negation word is required. Note that the examples given in the right-hand column are incorrect English.

	MORE COMMON / LESS FORMAL	LESS COMMON / MORE FORMAL (1), NO (2)
1	There are **not many** options available.	There are **few** options available.
1	We **don't** have **much** time available.	We have **little** time available.
1	There are **not as many** opportunities for women as there are for men.	There are **fewer** opportunities for women than for men.
1	There are **not** many cases where patients have such symptoms.	The cases where patients have such symptoms are **rare**.
2	The device was **not** designed to be connected to a network, **either** wired **or** wireless.	The device was **not** designed to be connected to a network, **neither** wired **nor** wireless.
2	The authors did **not** write **anything** regarding …	The authors did **not** write **nothing** regarding …

16 Word order: nouns and verbs

16.1 put the subject before the verb and as near as possible to the beginning of the phrase

1. Put the subject before the verb (for exceptions see 16.5, 16.6).
2. The subject generally contains the most important information. Put it as near as possible to the beginning of the sentence.

	YES	NO
1	**The referees' reports** have arrived.	They have arrived **the referees' reports**.
1	The **method** is important.	It is important the **method**.
2	**Several techniques** can be used to address this problem.	To address this problem **several techniques** can be used.
2	**Time and cost** are among the factors that influence the choice of parameters.	Among the factors that influence the choice of parameters are **time and cost**.
2	Although **algorithms** for this kind of processing are reported in the above references, the execution of …	Although in the above references one can find **algorithms** for this kind of processing, the execution of …

16.2 decide what to put first in a sentence: alternatives

1. Put the most important idea first, this will make your writing more direct and memorable.

	NORMAL POSITION OF THE SUBJECT	TO GIVE PARTICULAR EMPHASIS
1	**Lee** noted that 40% of the data was erroneous, contrasting with Hall's estimation of 20%.	Up to 40% of the data was misleading, **Lee** notes.
	Author stressed to contrast with other author.	Here the quantity is stressed.
1	A new cure was discovered recently [12].	It was only recently that a new cure was discovered [12].
		The time reference (*only recently*) gives interesting or surprising new information.

A. Wallwork, *English for Academic Research: Grammar, Usage and Style*, DOI 10.1007/978-1-4614-1593-0_16, © Springer Science+Business Media New York 2013

16.3 do not delay the subject

1. Put the subject first before mentioning when, how, where and why it functions. If you begin with a subordinate clause, this will force the reader to wait in order to find out what you are really referring to.
2. If you are using a lot of link words (13), such as *in particular, generally speaking, consequently, in addition*, don't always put them at the beginning. If possible, find a short word (*thus, so, also*) and insert it before the verb.
3. Avoid using an impersonal *it* at the beginning of the sentence. Instead use modal verbs (*might, need, should* etc.) or an adverb.
4. Avoid beginning a phrase with a time period containing the verb *to be.*

	YES	NOT RECOMMENDED (1–3) WRONG (4)
1	The **samples** were dried after they had spent five minutes in an aqueous solution, and 20 minutes in the cold room.	After five minutes in an aqueous solution, and a further 20 minutes in the cold room, the **samples** were dried.
1	Despite **Iceland's** favorable geological situation in terms of harnessing all kinds of geothermal resources, until a few years ago only geothermal-electric generation received much attention.	Despite **its** favorable geological situation in terms of harnessing all kinds of geothermal resources, until a few years ago only geothermal-electric generation received much attention in **Iceland**.
2	**The old system** should **thus** not be used.	**For this reason**, it is not a good idea to use the **old system**.
3	Users **should** be distributed evenly.	**It is recommended** to distribute users evenly.
3	This **can** be done with the new system.	**It is possible** do this with the new system.
4	We have been studying this problem **for three years.**	**They are three years** that we study this problem.
	= **For three years** we have been studying this problem and we still have no results.	**It is since three years** that we study this problem.

16.4 avoid long subjects that delay the main verb

Make sure the verb is near the beginning of the sentence and next to the subject. If the subject is very long, the reader will be left waiting to know what the verb is. To avoid this problem:

1. Use an active verb, rather than the passive form (10.3).
2. Shift the verb to the beginning of the sentence. This may involve changing the verb and / or changing the word order.
3. Divide up a long sentence into two shorter sentences.

	YES	NOT RECOMMENDED
1	**ABC generally employs** people with a high rate of intelligence, a proven talent for problem-solving, a passion for computers, along with good communication skills.	People with a high rate of intelligence, a proven talent for problem-solving, a passion for computers, along with good communication skills **are generally employed** by ABC.
2	This data shows that **there are** significant correlations between ...	This data shows that significant correlations between the cost and the time, the time and the energy required, and the cost and the age of the system **exist.**
2	Fonts can be easily configured **as well as** filters, ticker settings, blotters, and message bars.	Fonts, filters, ticker settings, blotters, and message bars **can easily be configured**.
3	People with a high rate of intelligence **are generally employed** by ABC. They must also have other skills including: a proven talent for problem-solving ...	People with a high rate of intelligence, a proven talent for problem-solving, a passion for computers, along with good communication skills **are generally employed** by ABC.

16.5 inversion of subject and verb

1. In questions containing the verb *to be*, auxiliary verbs (*have, had, will, would*), or modal verbs, invert the subject and verb.
2. Treat *to have* like a normal verb.
3. Be careful not to invert subject and verb after *what, which, who, where, why* when these are not used in a question.

	YES	NO
1	**Are doctors** becoming the new drug representatives? **Can we** allow them to have this role? How long **has this situation** been going on? **Would it** be right to intervene?	**Doctors are** becoming the new drug representatives? **We can** allow them to have this role? How long **this situation has** been going on? **It would** be right to intervene?
2	**Do we have** the resources to educate all children?	**Have we** the resources to educate all children?
3	We were unable to identify **what the problem was**.	We were unable to identify **what was the problem.**
3	The authors did not state **where their data came from**.	The authors did not state **where did their data come from**.

16.6 inversion of subject and verb with *only, rarely, seldom* etc.

1. If you put *only* or an adverb of frequency that indicates that an event almost never takes place (*rarely, seldom*) as the first word of a phrase, then you must invert subject and object as if you were forming a question (YES column below).
2. The same rule applies when you put a negation (e.g. *never, nothing*) as the first word in a phrase.

This construction is difficult to remember, so it is probably best to avoid it. Use the normal word order instead (third column).

	YES	NO	YES (ALTERNATIVE)
1	**Rarely does this happen** when the user is online.	**Rarely this happens** when the user is online.	This **rarely** happens when the user is online.
1	**Only when** all the samples have been cleaned, **can you** proceed with the tests.	**Only when** all the samples have been cleaned, **you can** proceed with the tests.	**You can only** proceed with the tests when all the samples have been cleaned.
2	**Never** before **had we** seen such a powerful reaction.	**Never** before **we had** seen such a powerful reaction.	**We had never** seen such a powerful reaction **before**.
2	**Not** just by overeating, but through lack of exercise, **do people become** overweight.	**Not** just by overeating, but through lack of exercise, **people become** overweight.	**People become** overweight through lack of exercise, not exclusively from overeating.

16.7 inversions with *so, neither, nor*

The subject and auxiliary are inverted after *so* and *neither* / *nor* when these are used to compare two or more items

1. *So* is used when the sentence is affirmative.
2. N*either* and *nor* have identical meanings and are used when the sentence is negative.

This construction is difficult to remember, so it is probably best to avoid it. Use the normal word order instead (third column).

	YES	NO	YES (ALTERNATIVE)
1	We found that helium is lighter than air, and so did Smith et al [2014].	We found that helium is lighter than air, and **also** Smith et al [2014].	**In line with** Smith et al [2014], we found that helium is lighter than air.
2	The alarm did **not** function and **neither did** the back up system.	The alarm did **not** function, **neither** the back up system.	The alarm did not function, **moreover** the back up system failed.

16.8 put direct object before indirect object

The direct object is the thing given or received. The indirect object (in bold in the table below) is the thing that the direct object is given to or received by. Look at the position of the *direct object* and indirect object in this sentence: "The authors sent *their manuscript* to the journal." Thus, the normal word order is: (1) subject (*the authors*), (2) verb (*sent*), (3) direct object (*their manuscript*), (4) preposition (*to*), (5) indirect object (*the journal*).

1. The kind of construction outlined above is often found with verbs followed by *to* and *with.*

Examples: *associate X with Y, apply X to Y, attribute X to Y, consign X to Y, give X to Y (or give Y X), introduce X to Y, send X to Y* (or *send Y X*)

2. If the direct object is very long and consists of a series of items, you can put the indirect object after the first item and then use *along with.*
3. As an alternative to rule 2, you can use a colon to introduce a list.
4. Not all sciences respect rule 1, and particularly in mathematical sciences you may find the indirect object before the direct object.

	YES	NO
1	We can separate P and Q with **this tool**.	We can separate, with **this tool**, P and Q.
	With **this tool** we can separate P and Q.	
1	We can associate a high cost with **these values**.	We can associate with **these values** a high cost.
2	We can associate a high cost with **these values**, along with higher overheads, a significant increase in man hours and several other problems.	We can associate with **these values** a high cost, higher overheads, a significant increase in man hours and several other problems
3	We can associate several factors with **these values**: a high cost, higher overheads, ...	
4	This is a rule that associates with **each element in S** a unique element in T.	

16.9 phrasal verbs

A simplified definition of a phrasal verb is a verb that is made up of one or more prepositions. Phrasal verbs tend not to be used in manuscripts as they are considered quite informal and more appropriate in the spoken language. Also, the same verb may have many different meanings, which could be confusing for the reader. However, some phrasal verbs are used in academia both in manuscripts, reports and emails e.g. *back up, break down, bring up, carry out, cut down, draw up, ease off, fall through, fill in, give off, go through, iron out, kick off, look forward to, phase out, point out, run into, set up, wear out*. Unfortunately there are different categories of phrasal verbs and by just looking at the verb it is impossible to know which category they belong to. Below are just two useful guidelines relating to the position of the direct object.

1. Some verbs require the pronoun to be inserted before the preposition.
2. Other verbs require the pronoun to be inserted after the preposition.
3. With some verbs you can put the direct object after the preposition or before. Separating the two parts so the verb (i.e. putting the object before the preposition) is more informal.

If you are not sure, the easiest solution is to keep the parts of the verb together, and avoid using pronouns and simply repeat the subject. Alternatively and where possible, use an alternative verb: e.g. *carry out (perform), cut down (reduce), go through (examine).*

	AFTER VERB	BETWEEN VERB AND PREPOSITION	NO
1	We **carried out the research**.	We **carried it out** in two stages.	We **carried out it** in two stages.
1	Smith **pointed this out** in his seminal paper.		Smith **pointed out this** in his seminal paper.
2	We **came across your paper** by chance.		We **came it across** by chance.
	We **came across it** by chance.		
3	We have **set up a new project**.	We have **set a new project up**.	We have **set up it.**
3	Smith **pointed out this fact** in his seminal paper.	Smith **pointed this fact out** in his seminal paper.	Smith **pointed out it.**

16.10 *noun + noun* and *noun + of + noun* constructions

1. In some cases, you can use either a *noun of noun* construction (e.g. *the University of Manchester*), or a *noun + noun* construction (e.g. *Manchester University*). Unfortunately there is no rule to help you decide if they are both applicable and if they both have the same meaning (2.4) For rules on when *'s* should be used, see genitive (2).
2. The *noun + of + noun* construction is generally used with words such as *piece, series, bunch, group* and *herd* (*flock* etc.).
3. In some cases, the *noun + of + noun* construction is not possible at all. This is often because the preposition *of* indicates that the first noun is made of the second noun (e.g. *a ring of gold = a ring made of gold*).
4. Long strings of nouns and adjectives are generally only used if they are names of pieces of equipment or methods.

	YES	NO
1	**Methods of** payment / **Payment** methods	**Payment's** methods
1	A **law of nature**	A **nature law** / A **nature's law**
1	A **software program** and a **hardware device**	A program of software and a **device of hardware**
1	Title: **Syringe exchange** and **risk of infection**	Title: The **exchange of syringes** and **risk infection**
2	The **series of plugs** was used together with two **groups of switches** and an innovative **piece of electrical equipment**.	The **plug series** was used together with two **switch groups** and an innovative **electrical equipment piece**.
3	A **shoe shop**	A **shop of shoes**
4	A recently developed reverse Monte Carlo quantification method	
4	A Hitachi S3500N environmental scanning electron microscope	

16.11 strings of nouns: use prepositions where possible

1. Do not put nouns in strings when the reader is unlikely to be able to understand how they relate to each other. Use prepositions to make the meaning clearer. This is particularly important in titles of papers – if the reader cannot understand your title then they will probably not read the paper.

2. A noun string can often be broken up by using a preposition: *of* = which belongs to, *for* = for the purpose of, *by* = how something is done and where necessary converting the nouns into verbs. This helps to clarify the relationships between the various nouns.

	YES	NO
1,2	Least Toxic Methods **for** Pest Control	Least Toxic Pest Control Methods
		Pest Control Least Toxic Methods
1,2	Quantifying surface damage **by measuring** the mechanical strength of silicon wafers.	Silicon wafer mechanical strength **measurement** for surface damage quantification.
2	The streets **of** San Francisco.	San Francisco streets
		San Francisco's streets
2	For **reasons of space**, we will not consider ...	For **space reasons**, we will not consider ...
2	Instructions **for boiling** potatoes	Potato boiling instructions

16.12 deciding which noun to put first in strings of nouns

Unfortunately there are no clear rules regarding which noun should go first. Also, the convention varies from discipline to discipline. For the rules on when to use an *'s* after the first noun, see genitive (2)

1. In many cases the first noun acts as an adjective that describes the second noun. In such cases the generic noun will normally go in second position.
2. When talking about families it depends on whether you are talking about people or insects, flowers etc. With human families use *surname + family*, in entomology, botany etc. use *family + species.* This is just one example illustrating the rather random nature of some aspects of the English language!

	YES	ALSO POSSIBLE
1	Press the **Control key**.	
1	Use **Track Changes** to make your revisions.	
2	More has been written about the **Kennedy family** than perhaps any other family in the history of the United States.	These mites are included in the **family *Tetranychidae***, Order Acarina, Class Arachnida.

16.13 position of prepositions with *which, who* and *where*

There are two possible positions for a preposition that is being used in conjunction with *which, who* and *where*:

1. Directly before *which, who* and *where* – this is a formal style and may sound strange. Note the use of *whom* (*with whom, from whom*).
2. At the end of the phrase – more informal and more common.
3. If there is already a preposition in the phrase, then the preposition associated with *who* or *which* is located at the end of the phrase.
4. B*y* cannot be separated from *which.*

	DIRECTLY BEFORE WHO / WHICH	AT END OF PHRASE
1,2	We want to know **to which** group the member belongs.	We want to know **which** group the member belongs **to**.
1,2	We want to know **from where** he comes.	We want to know **where** he comes **from**.
1,2	These were researchers **with whom** we had worked before.	These were researchers **who** we had worked **with** before.
1,2	Interviewees mark all the statements **with which** they agree.	Interviewees mark all the statements **which / that** they agree **with**.
1,2	The clinical symptoms of the children **from whom** the virus was isolated were similar to those found in adults.	The clinical symptoms of the children **who** the virus was isolated **from** were similar to those found in adults.
	In this case the position of *from* is appropriate.	Too informal for a manuscript.
3		This *depends on* **which** group the member belongs **to**.
4	The means **by which** the ER environment, is regulated have yet to be elucidated.	

17 Word order: adverbs

17.1 frequency + *also, only, just, already*

Adverbs of frequency (e.g. *always, sometimes, occasionally*) and words like *also, just, already,* and *only,* are usually placed:

1. Immediately <u>before</u> the main verb.
2. Immediately <u>before</u> the second auxiliary when there are two auxiliaries.
3. <u>After</u> the present and past tenses of *to be.*
4. For special emphasis, some adverbs (*sometimes, occasionally, often, normally, usually*) can be located at the beginning of a sentence.
5. When *only* is associated with a noun rather than a verb, it is located before the noun. It can also appear at the end of a sentence, but this is rare.

	YES	NO (* = POSSIBLE, BUT NOT COMMON)
1	You **only / also / just** need to sign the document.	You need **only / also / just** to sign the document.
1	We don't **usually** go abroad on holiday.	We **usually** don't go abroad on holiday.*
2	We would **never** have seen him otherwise.	We **never** would have seen him otherwise. *
2	This may not **always** have been the case.	This may not have been **always** the case.
3	They are **always** late in sending their manuscripts to the editor.	They are late **always** in sending their manuscripts to the editor.
4	**Normally** X is used to do Y, but **occasionally** X can be used to do Z.	
5	**Only** Emma has been to Japan. No one else has been to Japan. Emma has **only** been to Japan. But she has not been to China or Korea.	

A. Wallwork, *English for Academic Research: Grammar, Usage and Style,*
DOI 10.1007/978-1-4614-1593-0_17, © Springer Science+Business Media New York 2013

17.2 probability

Adverbs of probability (e.g. *probably, certainly, definitely*) go immediately before the:

1. Main verb.
2. Negation (*not* and contractions e.g. *don't, won't, hasn't*).

	YES	NO * OR NOT COMMON
1	She will **certainly** come.	She **certainly** will come.
		She will not come **certainly.***
2	She will **probably not** come.	She **probably** will **not** come.
	She **probably won't** come.	She will not **probably** come.*
		She will **not** come **probably.***
2	She **definitely hasn't** read it.	She **hasn't definitely** read it.

17.3 manner

An adverb of manner describes how something is done (e.g. *quickly*), or to what extent (e.g. *completely*). Some adverbs of manner can go before the verb. But, since <u>all</u> adverbs of manner can <u>always</u> also go after the verb or noun, it is best to put them there. You will then avoid mistakes.

1. Subject + verb + adverb of manner + full stop (.).
2. Subject + verb + noun + adverb [+ rest of phrase].

	YES	NO
1	This program could help **considerably**.	This program could **considerably** help.
2	This program will help system administrators **considerably**.	This program will help **considerably** system administrators.
	This program will help system administrators **considerably** to do x, y and z.	This program will help **considerably** system administrators to do x, y and z.

17.4 time

Adverbs of time:

1. Usually go at the end of the phrase, particularly if they consist of more than one word.
2. When used in contrast with each other, they go at the end.
3. In some cases (e.g. *today, tomorrow, tomorrow evening*) they can go at the beginning for emphasis.

	YES	NO
1	We will go there **once or twice a week / as soon as possible**.	**Once or twice a week / as soon as possible** we will go there.
1	We will go there **immediately**.	We will **immediately** go there.
		We will go **immediately** there.
2	We will go there **tomorrow morning not tomorrow evening**.	**Tomorrow morning** we will go there **not tomorrow evening**.
3	**Today**, we are going to talk about the position of adverbs.	We **today** are going to talk about the position of adverbs.

17.5 *first(ly), second(ly)* etc.

When you are listing events:

1. Put the adverb (*firstly, secondly* etc.) at the beginning of the phrase. You can say *firstly* or *first, secondly, thirdly, fourthly* etc. are preferred to *second*, third, fourth etc. in a manuscript. *first* is usually followed by *then* rather than *secondly.*
2. *Then* can be placed at the beginning of the sentence, but is more common before the main verb.

	YES	NO
1	**First / Firstly**, we will do X. Then we will do Y. **Finally**, we will do Z.	We will **firstly** do X. Then we will do Y.
		We will **finally** do Z.
2	Initially, we used X. **Then** we decided to use Y.	
	At the beginning we used X, we **then** decided to use Y.	

17.6 adverbs with more than one meaning

There are a few adverbs that change meaning depend on their position (i.e. before or after the verb):

1. **Normally**: before = *usually*, after = the opposite of *abnormally* (this usage is not very common, *in the normal way* is more common).
2. **Clearly**: before = *obviously*, after = *without difficulty.*
3. **Fairly**: before = *quite* (in a sufficient manner), after = *in the right proportion.*

	BEFORE THE VERB	AFTER THE VERB
1	Patients **normally** undergo rehabilitation after such accidents.	After six months of rehabilitation 65% of the patients were able to walk **normally** (i.e. without assistance).
2	**Clearly**, the authors have not followed the instructions carefully.	The instructions were not written **clearly**, in fact they were almost impossible to understand
3	The article is written **fairly** well, but needs improving in several areas.	Profits were not distributed **fairly** amongst the shareholders, which led to several complaints.

17.7 shift the negation word (*no, not, nothing* etc.) to near the beginning of the phrase

Negations generally contain key information so they should be located as near as possible to the beginning of the sentence. By doing so, you signal to the reader that you are about to say something negative rather than something affirmative. It can be misleading to put the negation at the end. So put the following near the main verb:

1. *Not* and *no.*
2. Adverbs that contain negative information, for example: *only, rarely, seldom, never.*
3. Note the position of *or not* when associated with *whether* + verb.

	YES	WRONG (*) OR NOT OPTIMUM
1	This **did not** seem to be the case.	This seemed **not** to be the case.*
1	There is **almost no** documentation on this particular matter.	Documentation on this particular matter **is almost completely lacking**.
1	We did **not** find **anything** to contradict these results.	We found to contradict these results **nothing**.*
	= We found **nothing** to contradict these results.	
1	Finally, **no** noticeable post-copulatory behaviour was observed in this species.	Finally, a noticeable post-copulatory behaviour was **not** observed in this species.
1	The referees **did not find** the results interesting.	The referees found the results **not** interesting.*
1	Our results revealed that there is **no** relationship between X and Y.	Our results revealed that a relationship between X and Y does **not** exist.
2	This **rarely** happens when the user is online.	The number of times this happens when the user is online is generally **very few**.*
		The frequency of this event when the user is online is **rare**.*
2	We **only** realized this at the end of the tests.	We realized this **only** at the end of the tests.
3	This study investigates the influences affecting a physician's decision **whether or not** to prescribe medicines.	This study investigates the influences affecting a physician's decision **whether** to prescribe **or not** medicines.*
	=decision **whether** to prescribe medicines **or not**.	

18 Word order: adjectives and past participles

18.1 adjectives

Adjectives generally go before the noun they describe. An adjective often contains information that is more important than the noun it describes, because the adjective helps to discriminate between two different types of the same noun e.g. He has a *red* car, I have a *blue* car.

1. Put the adjective before the noun it describes.
2. If you put the adjective after the noun, then precede the adjective with *that / which / who + verb.*
3. An exception to Rule 3 is with *available* and *possible*, which are often found after the noun.
4. Adjectives are not usually found between two nouns.
5. Do not put an adjective before a noun that it does not describe.

	YES	NO
1	This is a **good and interesting** book.	This is a book **good and interesting**.
1	He is an **intelligent** student.	He is a student **intelligent**.
2	He is a **student who is intelligent** enough to pass the exam.	He is a **student intelligent** enough to pass the exam.
3	The software **available** does not solve this problem.	The **available** software does not solve this problem.
3	This appears to be the only solution **possible / possible** solution.	
4	The **main** features of the software.	The software **main** features.
4	The **computational** complexity of the algorithm.	The algorithm **computational** complexity.
5	The **main contribution** of the document.	The **main document** contribution.

A. Wallwork, *English for Academic Research: Grammar, Usage and Style*,
DOI 10.1007/978-1-4614-1593-0_18, © Springer Science+Business Media New York 2013

18.2 multiple adjectives

1. A very general guideline for a string of adjectives is: size + age + color + origin + material + use.
2. When deciding the order, first choose the main adjective (or noun acting as an adjective) which is typically found with the associated noun e.g. *software solutions*. Then precede with a maximum of three more adjectives e.g. *an extremely effective (and) innovative software solution*. In this example both *effective* and *innovative* have a similar function and are interchangeable. *extremely* relates to both *effective* and *innovative* and must therefore go before these two adjectives.
3. Adjectives are located after past participles.
4. The position of the adjective can change the meaning of the phrase.
5. To aid clarity, consider changing an adjective into a noun, and modifying the word order.

	YES	NO
1	His swimming costume, which was large, old and red, was made in England and from cotton. It was found in … = His large old red English cotton swimming costume was found in …	A red old English cotton large swimming costume.
2	The **low stock size** of **edible Asian species** has led to the need for new resources overseas.	The **stock low size** of **Asian edible species** has led to the need for new resources overseas.
2	All the **ready-to-eat jellyfish products** that were examined had been contaminated.	All the **examined jellyfish ready-to-eat products** had been contaminated.
2	The **mean daily air temperature** was measured.	The **mean air daily temperature** was measured.
3	They were **colored red and white**.	They were **red and white colored**.
4	The **female's first choice** was … There is only one female involved. The **first female's choice** was … This implies that there was at least a second female involved.	A variety of choices were offered both to the male and the female. Interestingly, the **first female's choice** was …
5	Products sold in Chinese communities **in France**.	Products sold in **French** Chinese communities.

18.3 ensure it is clear which noun an adjective refers to

If an adjective is followed by two nouns, it may not be clear to the reader if the adjective only refers to the first noun, or both to the first and second nouns. If there could be ambiguity, then you need to rearrange the phrase:

1. If the adjective (e.g. *new*) only refers to the first noun (e.g. *teachers*) either (1) change the order of the nouns, or give each noun a different adjective.
2. If the adjective refers to both of the nouns, and if you think there could be ambiguity, then (1) put the adjective before both nouns, or (2) rearrange the sentence.

	YES	NO
1	The course is intended for students and **new teachers**.	The course is intended for **new teachers and students**.
	= The course is intended for **new teachers** and **all students**.	
2	The course is intended only for **new** teachers and **new** students.	The course is intended for **new teachers and students**.
	= The course is intended only for **newcomers**: **both** teachers and students.	

18.4 past participles

1. In most cases past participles can always go after the noun, but in many cases they cannot go before. So, put them after and you will probably be right!
2. In some cases both positions are possible, though when the past participle is located after the noun it is often followed by further details.
3. Be careful with *used*. Before the noun it means 'second hand', after the noun it means 'which is used'.

	YES	NO OR NOT COMMON
1	It shows details of all the **results found**.	It shows details of all the **found results**.
1	The **data reported** show that …	The **reported data** show that …
1	We detail the main **social actors involved** along with all the **materials consumed**.	We detail the main **involved social actors** along with all the **consumed materials**.
1	The **alternatives considered** and the way the problem is structured may vary in interpretation.	The **considered alternatives** and the way the problem is structured may vary in interpretation.
2	It shows details of all the **specified actions**.	
	It shows details of all the **actions specified** (in the manual).	
2,3	This was the **application used** by the testers.	This was the **used application** by the testers.
3	I bought a **used car**.	
	i.e. a second-hand car	

19 Comparative and superlative: *-er*, *-est*, irregular forms

19.1 form and usage

1. All monosyllable adjectives require *-er / -est* (exceptions: *more true* or *truer*, *more real*). All adjectives with three or more syllables require *more / most.*
2. Two-syllable adjectives ending in a vowel sound (e.g. *easy*, *happy*, *narrow*) take *-er / -est*, whereas those ending in a consonant sound (e.g. *complex, massive, useful*) require *more / most*. Note: *clever, common, friendly, gentle, quiet* and *simple* and be used with either form *(most common, commonest).* See also spelling (28.1).
3. Use the comparative form (e.g. *bigger, better, more beautiful*) to compare two things or two groups of things.

	YES	NO
1	This is the **biggest** and **most productive** machine in the world.	This is the most **big** and **productivest** machine in the world.
2	This is the **busiest** and **heaviest** period of the year, but yet also the **most peaceful**.	This is the **most busy** and **most heavy** period of the year, but yet also the **peacefullest**.
3	Brazil is **bigger** than Argentina.	Brazil is **biggest** than Argentina.
3	The system performed **better / worse / less efficiently / more efficiently** in the first test than in the second test.	The system performed **best / worst / least efficiently / most efficiently** in the first test than in the second test.

A. Wallwork, *English for Academic Research: Grammar, Usage and Style*,
DOI 10.1007/978-1-4614-1593-0_19, © Springer Science+Business Media New York 2013

19.1 form and usage (cont.)

4. Use the superlative form (e.g. the *biggest, the best, the most beautiful*) to describe something in absolute terms. Note that *the* is used before all superlatives, except for the case given in Rule 6.
5. Note these irregular forms: *good / better / best; bad / worse / worst; far / further / furthest* (alternative spelling: *farther / farthest*).
6. Note the difference between (a) *Poverty in London was the highest in England* (b) *Poverty was highest in England.* In (a) we are talking about two places that are in relation to each other (London and England). In b) we are talking about poverty without putting two countries in direct relation to each other. This subtle difference is only applicable when the superlative does not directly precede a noun.

	YES	NO
4	The application returns only the **most relevant** results.	The application returns only the **more relevant** results.
4	It always chooses the **best** solution.	It always chooses the **better** solution.
4	Mumbai and Sao Paulo are big cities, but Tokyo is **the biggest** and **most populated** in the world.	Mumbai and Sao Paulo are big cities, but Tokyo is **the bigger** and **more populated** in the world.
4	This candidate was certainly **the best**.	This candidate was certainly **best**.
5	They traveled **further** than the others.	They traveled **farer** than the others.
6	Production was **lowest** among IT companies.	Production was **the lowest** among IT companies.
	= **The lowest values** of production were achieved by IT companies.	
6	Mortality / Obesity / Reliability / Efficiency / Concentration was **highest** in / among / for patients diagnosed with …	
	= **The highest levels** of mortality …	

19.2 position

1. Place comparatives and superlatives before the noun they describe.
2. If you need to put a comparative after the noun, then precede it with *that*.

	YES	NO
1	This solution has **more serious** drawbacks than the other solution.	This solution has drawbacks **more serious** than the other solution.
2	The application returns only the **results that are the most relevant**.	The application returns only the **results most relevant**.

19.3 comparisons of (in)equality

1. Use *than* when comparing two or more items. Avoid unnecessary use of *with respect to / in comparison to / compared to.*
2. To say that items are the same, use *the same as.*
3. When indicating that two things are equal in terms of a particular quality, use *as … as.*
4. When making negative comparisons *less* tends to be used only with multi-syllable adjectives. Use *not as … as* with monosyllables or multisyllables.

	YES	NO
1	China is bigger **than** the United States.	China is **bigger of** the United States.
		China is **big with respect to** the USA.
2	Australia is approximately **the same** size **as** the 48 mainland states of the USA.	Australia is approximately the same size **than / of** the 48 mainland states of the USA.
3	This book is **as good / expensive as** that book.	This book is as good / expensive **than** that book.
4	This solution **is not as good as** the other one.	This solution is **less good than** the other one.
4	The first is not as good **as** the second.	The first is not so good **like** the second.
4	This solution is **not as efficient as** the other one.	This solution is less efficient **as** the other one.
	= This solution is **less efficient than** the other one.	

19.4 *the more ... the more*

1. The verb is placed after the subject and not before.
2. The definite article (*the*) is required before each comparative.
3. On some occasions, no verb is required.

	YES	NO
1	In realistic conditions, the more **robust the software is**, the less problems there are.	In realistic conditions, the more **is robust the software**, the less problems there are.
2	**The more** you use the software, **the easier** it becomes.	**More** you use the software, **easier** it becomes.
3	**The sooner** the job is done, **the better**	**The sooner** the job is done, **better is**

20 Measurements: abbreviations, symbols, use of articles

WRITTEN	SAID	WRITTEN	SAID
CARDINALS AND ORDINALS			
101	a / one hundred and one	58,679	fifty eight thousand six hundred and seventy nine
213	two hundred and thirteen	2,130,362	two million, one hundred and thirty thousand, three hundred and sixty two
1,123	one thousand, one hundred and twenty three		
13th	thirteenth	31st	thirty first
CALENDAR DATES			
10.03.20	the tenth of March two thousand and twenty (GB)	1996	nineteen ninety six
			nineteen hundred and ninety six
GB: day / month / year	or March (the) tenth two thousand and twenty (GB)	1701	seventeen oh one
			seventeen hundred and one.
US: month / day / year	October third two thousand twenty (US)	2010s	twenty tens
FRACTIONS, DECIMALS, PERCENTAGES			
¼	a quarter / one quarter	0.25	(zero) point two five
½	a half / one half	0.056	(zero) point zero five six
¾	three quarters	37.9	thirty seven point nine
10%	ten per cent	100%	one hundred percent

(continued)

A. Wallwork, *English for Academic Research: Grammar, Usage and Style*,
DOI 10.1007/978-1-4614-1593-0_20,

(continued)

WRITTEN	SAID	WRITTEN	SAID
SQUARES, CUBES ETC.			
4 m^2	four meters squared, four square meters	2^5	two to the power of five
5 m^3	five cubic meters, five meters cubed		
MONEY			
678	six hundred and seventy eight euros	$450,617	four hundred fifty thousand six hundred seventeen dollars
¥1.50	one yen fifty (cents)	$1.90	a dollar ninety
MEASUREMENTS			
1 m 70	one meter seventy	3.5 kg	three point five kilos
3 m × 6 m	three meters by six		
100^0	one hundred degrees	-10^0	minus ten degrees
			ten degrees below zero
PHONE NUMBERS			
0044 161 980 4166	zero zero four four one six one nine eight zero four one double six	ext. 219	extension two one nine
	or oh oh four four etc.		

20.1 abbreviations and symbols: general rules

1. The humanities and social sciences tend to use words rather than abbreviations and symbols.
2. Symbols generally come after the number. Exceptions: currencies (e.g. ¥100, €56).
3. Numbers before abbreviations and symbols must be digits (e.g. 7) rather than words (e.g. *seven*).
4. Abbreviations for measurements are not usually followed by a period (.) unless at the end of a sentence. They do have a plural form.
5. Most abbreviations for measurements are all lower case. Exceptions: bytes (e.g. GB, KB); micro measurements (mL – microliter, milliliter); and temperatures (C, F).
6. Abbreviations in a series tend to be repeated.
7. In a range, the abbreviation tends to go with the last item.
8. Do not use an abbreviation for a measurement without a number.

	YES	NO
1	It took King Harold's men **ten days** to cover the **400 kilometers** from York to fight at the battle of Hastings in temperatures that ranged from **twenty degrees below zero to three degrees above**.	It took King Harold's men **10 d** to cover the **400 km** from York to at the battle of Hastings in temperatures ranging from **−20°C to 3°C**.
2	The total cost was **$5000**.	The total cost was **5000$**.
	= ... was **5000 USD / US dollars**	
3	The patient weighed **65kg**.	The patient weighed **sixty five kg**.
4	The patient weighed **65kg** and was **120 cm** tall.	The patient weighed **65 kgs** and was **1.20 cm.** tall.
5	The patient weighed **65kg**.	The patient weighed 65**Kg**.
5	A memory of **3 GB**.	A memory of **3gb**.
6	The three patients weighed **65kg, 75kg and 85kg**.	
7	... from **65 to 85kg**.	
8	A few **micrograms** (e.g. **3μg**) ...	A few **μg** of (e.g. **3μg**) ...

20.2 spaces with symbols and abbreviations

1. There do not seem to be fixed rules about whether to put spaces before units of measurement. Check with your journal's style.
2. If the unit of measurement would appear alone at the beginning of the next line, then remove the space.
3. When describing computer memory, the style is generally not to use a space.
4. Do not insert a space between a number and *st, rd* and *th.*

	YES	NO
1	The patient weighed **65 kg / 65kg**.	
2	These rocks weighed up to **165kg** each.	These rocks weighed up to **165** **kg** each.
2	The temperature was **– 20°C**.	This meant that the temperature was **– 20°C**.
3	A **120GB** memory.	A **120 GB** memory.
4	He was born on March **10th**.	He was born on March **10 th**.

20.3 use of articles: *a / an* versus *the*

1. Use *a / an* to relate one unit of measurement to another.
2. Use *the* in measurements that begin with *by.*
3. Use *a / an* with *speed, rate* etc. when such words are followed by a number.
4. Use *the* with *speed, rate* etc. when such words are followed by a noun.

	A / AN	THE
1, 2	Gold may soon cost **$2000 an ounce**.	Gold is sold **by the ounce**.
3, 4	The disc gyrates at **a speed of 45 rpm**.	The pulses travel outward at **the speed of sound.**

20.4 expressing measurements: adjectives, nouns and verbs

Measurements using nouns can normally be expressed in several ways, either with the verb *to have* or the verb *to be*.

1. When the measurement appears after the noun, then use *of* as the preposition (*a width of 2 cm*); if it appears before the noun use *in* (*2 cm in width*).
2. Adjectives can be used instead of nouns.
3. The adjective can appear before or after the noun. Note the use of hyphens (25.6).
4. When measurements appear in brackets they are often not written as full sentences.

	YES	ALTERNATIVE
1	The **length of** the field **was** 200 meters. The field **was** 200 meters **in length**.	The field **had** a **length of** 200 meters.
1	These cores **were** approximately **1.5 mm in diameter** and **25 mm in height**.	These cores **had** an approximate **diameter of 1.5 mm** and a **height of 25 mm**.
2	The girl **was** 120 cm **tall**.	The girl **had** a **height** of 120 cm.
2	A is **as wide as** B.	A is **the same width as** B.
3	It was a **200-meter-long** field.	The field was **200 meters long**.
4	Samples were individually stored in fresh glass vials (**diameter: 1 cm; length: 6 cm**) until the testing phase.	

21 Numbers: words versus numerals, plurals, use of articles, dates etc.

21.1 words versus numerals: basic rules

1. If a number has to appear at the beginning of a sentence use the word (e.g. *eleven*) rather than the numeral (e.g. 11).
2. If necessary, rearrange the sentence so that the number does not appear at the beginning.
3. If it is not possible to apply Rule 2, use words instead.

	YES	NO
1	**Two hundred** samples were examined.	**200** samples were examined.
2	This feature is not used by **50%** of users.	**50%** of users do not use this feature.
2	**An amount of 1.85 mL** of distilled water was added to the mixture.	**1.85 mL** of distilled water was added to the mixture.
3	**Seventy per cent** of managers believe that praising employees makes no difference to performance.	**70%** of managers believe that praising employees makes no difference to performance.

A. Wallwork, *English for Academic Research: Grammar, Usage and Style*,
DOI 10.1007/978-1-4614-1593-0_21,

21.2 words versus numerals: additional rules

1. When you use numbers from one to eleven within a written text, write them as words (e.g. *nine*) rather than numerals (e.g. *9*). The reason for this is visual: it is harder to see a digit in a text than a word, e.g. *1* is harder to see than *one*. Exceptions: 21.3.
2. Consider using words for numbers above ten if this will facilitate reading.
3. Do not mix words and digits to refer to the same number, unless this number is a million or more.
4. Do not mix words and digits within the same context.
5. Times of day are written as numerals; use the 24 hour clock to avoid having to use *a.m.* (before midday) or *p.m.* (after midday).

	YES	NO
1	For the color measurements, **three** fruits of each cultivar were analyzed.	For the color measurements, **3** fruits of each cultivar were analyzed.
2	Of the 270 examined faecal samples, 46 were positive for Trichuridae eggs: **six** (2.2%) were positive for *E. boehmi* (Fig. 1a), **twelve** (4.4%) *E. aerophilus* (Fig. 1b) and **thirty-three** (12.2%) for *T. vulpis* (Fig. 1c).	Of the 270 examined faecal samples, 46 were positive for Trichuridae eggs: **6** (2.2%) were positive for *E. boehmi* (Fig. 1a), **12** (4.4%) *E. aerophilus* (Fig. 1b) and **33** (12.2%) for *T. vulpis* (Fig. 1c).
2	In **Tables 1 and 2, twenty** samples with …	In Tables **1 and 2, 20** samples with …
3	There were **200,000** people at the conference.	There were **200 thousand** people at the conference.
	There were **two hundred thousand** people at the conference.	
3	More than half of the Earth's **7.4 billion** inhabitants live in the tropics and subtropics.	More than half of the Earth's **7,400,000,000** inhabitants live in the tropics and subtropics.
4	There were **two- to three**-fold increases.	There were **two- to 3-fold** increases.
5	Rats were fed at **09.00** and **17.00** every day.	Rats were fed at **9 o'clock** in the morning and at **5 p.m.** every day.

21.3 when 1–10 can be used as digits rather than words

1. Use digits not words when numbers are in association with percentages, abbreviations for measurements, tables, and figures etc.

You can optionally use digits rather than words when:

2. The second number in a range of numbers is higher than eleven; alternatively write both numbers as words.
3. There is a series of numbers, or in ratios and proportions.
4. Numbers act as adjectives. Note the use of hyphens (25.6).

	YES	ALSO POSSIBLE
1	As shown in **Table 3**, the patient was only **1.20 m** tall and weighed **9 kg**. Her percentage body fat was **9.9%.**	
2	The process usually takes between **4 and 12** days.	The process usually takes between **four and twelve** days.
3	In the last three years the numbers have risen by **11, 6 and 7**, respectively.	In the last three years the numbers have risen by **eleven, six and seven**, respectively.
3	Multiple mating by females occurred in only **5 out of 34** species.	Multiple mating by females occurred in only **five out of thirty-four** species.
4	a **3-point** turn, a **4-day** week, a **size 7** component, a **6-year-old** child	a **six-year-old** child

21.4 making numbers plural

1. Whole numbers do not require an *-s* to indicate the plural and no preposition is used between the number and the noun.
2. An exception to Rule 1 is in expressions such as *tens, dozens, hundreds, thousands*, i.e. to indicate large generic numbers. In such cases *of* follows the number.
3. Fractions require an *-s* plural.
4. For reasons for readability, make single digits plural using *'s*. However, for other numbers (including dates) simply add an *s*.
5. A noun which follows a number is used in the singular form when acting as an adjective (technically these are called 'numerical modifiers'). Note the use of hyphens (25.6).

	YES	NO
1	**Four thousand experiments** have been conducted so far.	**Four thousands of** experiments have been conducted so far.
2	**Hundreds of** people were at the conference.	**Hundred** of people were at the conference.
3	One and a half **hours** (= an hour and a half), three **quarters** of an hour, four **fifths** of a liter, nine **tenths** of a second.	One and a half **hour**, three **quarter** of an hour, four **fifth** of a liter, nine **tenth** of a second.
4	The table contains only **0's** and **1's**.	The table contains only **0s** and **1s**.
4	In the **1990s**, many airlines flew Boeing **747s.**	In the **1990's**, many airlines flew Boeing **747's.**
5	A 51-**year**-old patient	a 51 **years** old patient
	i.e. a patient who is 51 years old	
	multi-**megabyte** memory	multi **megabytes** memory

21.5 singular or plural with numbers

1. Numbers and quantities require the verb that follows to be in the singular form. This is because they are seen as a mass rather than individual items.
2. The use of *there is / was* and *there are / were* depends on whether the noun that follows is in the singular or plural, respectively.
3. Use *another*, not *other*, before a number.
4. *None* is followed by a verb in the plural.

	YES	NO
1	Two weeks **is** not enough.	Two weeks **are** too few.
1	Three hundred kilometers **is** not too far.	Three hundred kilometers **are** not too far.
1	Clearly, $1,000,000 **is** a lot of money.	Clearly, $1,000,000 **are** a lot of money.
2	In this diagram there **is** a rectangle and two squares. = In this diagram there are **two** rectangles and a square.	In this diagram there **are** a rectangle and a square.
3	We need to do **another three** tests.	We need to do **other three** tests.
	= We need to do **three other** tests.	
4	None of the tests **give** optimum results.	None of the tests **gives** optimum results.

21.6 abbreviations, symbols, percentages, fractions, and ordinals

1. Always use numerals with abbreviations or symbols. Do not combine spelled forms of numbers with symbols.
2. P*ercentage* is one word, both *percent* (one word) and *per cent* (two words) are correct; do not use *%age.*
3. In ranges of percentages, either put the percentage symbol tends after the second number or after both numbers.
4. Fractions and ordinal numbers should not appear as digits (e.g. *1 / 4, 2nd*) at the beginnings of sentences or between other words.
5. Decimals are not written as words.
6. Decimals are written with a point (.) rather than a comma (,).
7. Commas tend to be used in whole numbers above 999 (but not in dates or horsepower).

	YES	NO
1	$2,000 / two thousand dollars	$two thousand
1	68c / sixty-eight cents	sixty-eight c
1,2	45% / forty-five per cent	forty-five%
1,2	The percentage of students who …	The %ge of students who …
3	The disease is fatal in **2–3%** of cases.	The disease is fatal in **2%–3** of cases.
	The disease is fatal in **2%–3%** of cases.	
4	**Two thirds** of those interviewed said that **one fifth** of their income was spent on fuel.	2/3 of those interviewed said that 1/5 of their income was spent on fuel.
4	The **first** and the **second** experiments proved the most successful.	The **1st** and **2nd** experiments proved the most successful.
5	The student scored **2.4** and **2.6** in the first two tests.	The student scored **two point four** and **two point six** in the first two tests.
6	The student scored **0.4** and **1.6** in the first two tests.	The student scored **0,4** and **1,6** in the first two tests.
7	The faculty has a total of **24,563** students.	The faculty has a total of **24563** students.

21.7 ranges of values and use of hyphens

You can introduce a range of values in three different ways:

*There should be **11–20** participants.*
*There should be **from 11 to 20** participants.*
*There should be **between 11 and 20** participants.*

Use a hyphen (25.6):

1. To indicate a range of values with numerals. But to indicate a range of values with words, use *to.*
2. With fractions that are made up of two words (e.g. *three-fifths, seven-ninths*).
3. With ages and periods of time. Note that there is no plural *s* on the time period.

	YES	NO
1	The courses last **15–20** weeks.	The courses last **fifteen–twenty weeks.**
1	The course will last **three to four** weeks.	The course will last **three–four** weeks.
2	**Three-quarters** of the employees in this institute come to work by car.	**Three quarters** of the employees in this institute come to work by car.
3	**Four-week** holidays can only be taken by **40-year-old** employees.	**Four weeks** holidays can only be taken by **40 years old** employees.

21.8 definite article (*the*) and zero article with numbers and measurements

Use the zero article (5):

1. With percentages and fractions.
2. Before the following words (and similar words) when they are followed by a number: *figure, appendix, table, schedule* etc.; *step, phase, stage* etc.; *question, issue, task* etc., *case, example, sample* etc.
3. With weights, distances etc.
4. In the expression *on average.*

Use the definite article (4):

5. Measurements that begin with *by.*
6. A number that has already been mentioned.

	YES	NO
1	Almost **80%** of scientific papers are published in English.	Almost **the 80%** of scientific papers are published in English.
1	More than **half** of the patients were infected with HIV.	More than **the half** of the patients were infected with HIV.
2	See the table in **Section** 2.	See the table in **the Section** 2.
2	We weighed **Sample 1** and **Sample 2** (see **Figure** 3).	We weighed **the Sample 1** and **the Sample 2** (see **the Figure 3**).
2	Details can be found in **Schedule** 2.	Details can be found in **the Schedule** 2.
3	The sample weighed **3 kg** / **three kilos**.	The sample weighed **the 3 kg / the three kilos**.
4	**On average**, debt rises by about $400 a month.	**On the average**, debt rises by about $400 a month.
5	Gold is sold by **the ounce** while coal sells by **the ton**.	Gold is sold by **ounce** while coal sells by **ton**.
6	Values must not go over a 90% threshold. This means that any values that go over **the 90%** threshold are not considered.	

21.9 definite article (*the*) and zero article with months, years, decades and centuries

1. Use the zero article (5) before months (e.g. *July, August*) and years (e.g. 1992, 2013, 2024).
2. Use the definite article (4) to refer to decades and centuries.

	YES	NO
1	Work began in **July** and is only expected to end in **2030**.	Work began in **the July** and is only expected to end in **the 2030**.
2	Research on this topic started **in the late 1990s**.	Research on this topic started **in late 1990s**.
2	**The twenty-first century / The 21st century** will witness the end of many minerals.	**Twenty-first century / 21st century** will witness the end of many minerals.
2	From **the 15th to the mid 16th century**, important changes were made to the techniques used in Chinese painting.	From **15th to mid 16th century**, important changes were made to the techniques used in Chinese painting.

21.10 *once, twice* versus *one time, two times*

1. *Once* = one time, *twice* = two times. *once* and *twice* are more commonly found than *one time* and *two times* – avoid mixing the two forms in the same phrase. *thrice* (three times) is archaic and should not be used.
2. *Once* and *twice* cannot be used after expressions such as a *minimum / maximum of.*

	YES	NO
1	The tests should be repeated at least **two or three times**.	The tests should be repeated at least **twice or three times**.
2	The test should repeated a **minimum of two times**.	The test should repeated a **minimum of twice**.

21.11 ordinal numbers, abbreviations and Roman numerals

There are three forms of writing ordinal numbers:

Form (A) word e.g. first, second, third, fourth.
Form (B) abbreviation e.g. *1st, 2nd, 3rd, 4th.*
Form (C) Roman numeral e.g. *I, II, III, IV.*

1. Use Form A within the main text of a manuscript.
2. Use Form B with centuries, millenniums, dynasties etc.
3. Form B can also be used (e.g. July 4th) but usage with the cardinal form is equally acceptable and avoids possible errors with -*st, -rd,* and *-th.*
4. Some conference names use Form A, others Form B, and others Form C – there appears to be no rationale for deciding which form to use. Note: do not mix the forms (e.g. *IIIrd*).
5. Use Form C with the names of people.
6. Arabic numerals (1, 2, 3) are used much more frequently than Roman numerals (I, II, III) for section numbering in papers.

	YES	NO (1–5), LESS COMMON (6)
1	This is the **first** time that … During the **third** experiment we …	This is the **1st** time that … During the **3rd** experiment we …
2	They can be dated to a time-span ranging from the **7th century** BC to the **2nd century** AD.	They can be dated to a time-span ranging from the **VII century** BC to the **II century** AD.
3	The Second Conference on Jugular Architecture will be held on **3 April 2026**.	The Second Conference on Jugular Architecture will be held on **3th** April 2026.
4	A summary of this paper was presented at the **Fourth / 4th / IV** Euroanalysis Conference, Helsinki.	A summary of this paper was presented at the **IVth** Euroanalysis Conference, Helsinki.
5	John Paul Getty **III**, King William **IV** and Pope John Paul **II** never met all together, but if they had …	John Paul Getty **3rd**, King William **4th** and Pope John Paul **2nd** never met all together, but if they had …
6	This is dealt with in more detail in **Sections 3 and 4**.	This is dealt with in more detail in **Sections III and IV**.

21.12 dates

1. Write centuries with Arabic numerals not Latin numerals. Note: non-religious alternatives to BC (before Christ) and AD (anno domini – year of our Lord) are BCE (before common era) and CE (common era), and also BPE (before present era) and PE (present era). However these acronyms are not, as yet, very common.
2. Write decades in their full numerical form (1980s) rather than abbreviated form ('80s) as otherwise there could be confusion between centuries. Also use the plural *s* without an apostrophe.
3. Write the first decade of each century in words not numerals. Note that 2000s could refer to the period from 2000-2009, or 2000-2099.
4. The world has three principal systems for writing dates:

   Form A: (dmy) 10 March 2020 = 10.03.2020.
   Form B: (mdy) March 10, 2020 = 03.10.2020.
   Form C: (ymd) 2020 March 10 = 2020-10-03.

The first form (number month year) is perhaps the clearest. To avoid confusion, always write the date with the month as a word.

	YES	NO
1	They can be dated to a time-span ranging from the **7th century** BCE to the **2nd century** CE.	They can be dated to a time-span ranging from the **VII century** BCE to the **II century** CE.
2	This paper presents an analysis of the techno-rhythms of the music of the **1990**s.	This paper presents an analysis of the music of the **'90s / 1990's**.
3	Little progress was made in the **first decade of the 21st century**, but considerable progress has been made in the **second decade / in the 2010s**.	Little progress was made in **2000s**, but considerable progress has been made in the **2010s**.
4	Smith et al. calculate that the world will end on **10 March 2030**.	They calculate that the world will end on **03.10.2030.** In the US, this would be interpreted as 10 October not 10 March.

22 Acronyms: usage, grammar, plurals, punctuation

22.1 main usage

1. The first time you use an acronym, write the word out in full, followed by the acronym in brackets. Afterwards, just use the acronym.
2. Each letter of an acronym is usually capitalized.
3. The written full form of an acronym may or may not require initial capital letters.
4. Exceptions to Rule 3 are frequent when one of the letters in the acronym is a preposition (typically *of*).
5. Acronyms that include a number can be found either in upper or lower case (e.g. *B2B* or *b2b* = *business to business*).
6. Do not repeat the final abbreviated word in the text following the abbreviation.

	YES	NO
1	Orders are dealt with on a **first in first out (FIFO)** basis.	Orders are dealt with on a **FIFO (first in first out)** basis.
2	We are part of a **NASA** project.	We are part of a **Nasa** project.
3	Users require a **p**ersonal **i**dentification **n**umber (PIN) to access any **N**orth **A**tlantic **T**reaty **O**rganization (NATO) files.	Users require a **P**ersonal **I**dentification **N**umber (PIN) to access any **n**orth **a**tlantic **t**reaty **o**rganization (NATO) files.
4	The future Internet is expected to support applications with quality of service **(QoS)** requirements.	The quality-of-service **(QOS)** requirements for …
5	Many **P2P / p2p** applications have now been blocked.	Many **peer2peer** applications have now been blocked.
6	The **GUI** is user friendly. It does not require a **PIN**.	The **GUI interface** is user friendly. It does not require a **PIN number**.

A. Wallwork, *English for Academic Research: Grammar, Usage and Style*, 191
DOI 10.1007/978-1-4614-1593-0_22, © Springer Science+Business Media New York 2013

22.2 foreign acronyms

Be careful of using acronyms that exist in your own language but not in English:

1. Put the meaning of the acronym before the acronym itself. If necessary also include the nationality.
2. If the acronym is unlikely to be unknown to your readers then it is helpful to give an explanation of what it stands for. This does not need to be a literal translation.
3. You do not need to explain the letters of an acronym in your own language when it has a clear English language translation or equivalent.
4. Ensure that you use the English form of an acronym when referring to an international entity, e.g. EU – *European Union*, not UE – *Union européenne.*
5. Even if capital letters are not used for an acronym in your language, they should be used in English - if not, they will give the appearance of a misspelled word.

	YES	NO
1	This paper describes a study by the **French National** Center for Scientific Research **(CNRS)** of …	This paper describes a **CNRS** (**National** Center for Scientific Research) study of …
2	Italian citizens are subject to various taxes, the most important being IRPEF, which is **a tax on personal income**.	Italian citizens are subject to various taxes, the most important being IRPEF **(Imposta sul Reddito delle Persone Fisiche – tax on the income of physical persons).**
3	The Brazilian ministry has control over the National Institute of Amazonian Research (INPA), and the National Institute of Technology (INT).	The Brazilian ministry has control over the National Institute of Amazonian Research (**Instituto Nacional de Pesquisas da Amazônia** – INPA), and the National Institute of Technology (**Instituto Nacional de Tecnologia** – INT).
4	The high commissioner of the **UN** stated that …	The high commissioner of **ONU / OOH** stated that …
5	Italian citizens are subject to various taxes, the most important being **IRPEF**.	Italian citizens are subject to various taxes, the most important being **Irpef**.

22.3 grammar

Acronyms, like all nouns, respect the normal rules of grammar:

1. If an acronym refers to a countable entity it requires an article when used in the singular.
2. The plural of an acronym is formed by adding an *s*.
3. If the last word in the full form of an acronym is plural, then a lower case *s* should be used at the end of the acronym. This rule does not apply to the names of some organizations (e.g. UN – United Nations).
4. Rule 3 applies even if the last letter in the acronym is an *s*. Note: in some cases, some authors choose to make the plural of an acronym that ends with an *s* by using *-es*; for example, computer scientists use both *ASs* and *ASes* as the acronym for *Autonomous Systems*.
5. Even though an acronym may have first been used in its singular form, this does not mean that it cannot then be used in the plural form. If an acronym is being used in a plural sense, then it must end in *s*.
6. Do not put an apostrophe before the plural form.

	YES	NO
1	We used **a PC**.	We used **PC**.
2	Four **PCs** in series were needed in order to make the calculation.	Four **PC** in series were needed in order to make the calculation.
3	This book is intended for non-native English teachers (hereafter **NNETs**).	This book is intended for non-native English teachers (hereafter **NNET**).
4	Solar **systems** (SS**s**) have been studied for thousands of years.	Solar systems (**SS**) have been studied for thousands of years.
4	Reactive oxygen **species** (ROS**s**) are important in a number of physiological and pathological processes.	Reactive oxygen species (**ROS**) are important in a number of physiological and pathological processes.
5	Enter your **PIN** (personal identification number). All users are required to have two **PINs**.	Enter your PIN (personal identification number). All users are required to have two **PIN**.
6	They released seven **CDs**.	They released seven **CD's**.

22.4 punctuation

1. Some acronyms have become words in their own right, and may be found with or without capitalization.
2. The letters of an acronym are not separated by periods (.). However, some authors write U.K. and U.S.A. rather than UK and USA.

	YES	ALSO POSSIBLE
1	We have developed a "what you see is what you get" (**WYSIWYG**) approach to map digitizing.	Following a **wysiwyg** philosophy, we have developed a novel approach to map digitizing.
1	The objections are part of a **NIMBY**, or Not in My Backyard, pattern of responses.	Policy scholars dedicated to efficient urban and industrial planning have long tried to understand the "**nimby** syndrome" in order to overcome local resistance to controversial land uses.
2	The **USA** and the **UK** are allies.	The **U.S.A.** and the **U.K.** are allies.

23 Abbreviations and Latin words: usage meaning, punctuation

23.1 usage

An abbreviation is the short form of word (example: *etc.* for *etcetera*).

1. Only use abbreviations for words such as *figure, table,* and *appendix*, when such words are associated with a number.
2. Abbreviations tend to be less readable, so consider only using them when you are short of space.
3. Don't use a percentage sign unless it is associated with a number.
4. Abbreviations for academic and other work positions are not generally found in manuscripts. Exception: *Dr* when *Dr* refers to someone with a PhD.
5. Abbreviations of academic degrees are not required when listing the names of the authors of your manuscript. For details on the abbreviations used for academic degrees in the UK and USA see http://en.wikipedia.org/wiki/British_degree_abbreviations; http://en.wikipedia.org/wiki/Academic_degree#Canada_and_United_States.

	YES	NOT RECOMMENDED
1	See the **figure** below.	See the **fig**. below.
2	See **Appendix** 1.	See **App**. 1.
2	See **Figure** 5 on **page** 10.	See **fig**. 5 on **p**. 10.
3	This value is always expressed as a **percentage**.	This value is always expressed as **%.**
4	These data were confirmed by **Professor** Lim, **Senator** Adams and **General** Kakowski.	These data were confirmed by **Prof.** Lim, **Sen.** Adams and **Gen.** Kakowski.
5	Psycholinguistics as a teaching aid J Win, A Yang, P Li	Psycholinguistics as a teaching aid J Win, **PhD**; A Yang, **EdD**; P Li, **MA**

A. Wallwork, *English for Academic Research: Grammar, Usage and Style*,
DOI 10.1007/978-1-4614-1593-0_23, © Springer Science+Business Media New York 2013

23.2 punctuation

1. Many authors use a capital letter with figure, table, appendix, both as full and abbreviated words. This rule only applies when these words are followed by a number.
2. The number that comes after an abbreviation is preceded by a space.
3. Abbreviations of quantities (examples: meters, kilograms) are not followed by a full stop (.). Write such abbreviations in lower case.

	YES	NO
1	See **Appendix** 1.	See **app** 1.
	See **App**. 1.	
2	See **Fig**. 1.	See **Fig**.1.
3	The road is **3 km** long.	The road is **3 km**. long.
		The road is **3 KM** long.

23.3 abbreviations found in bibliographies

Legend: ( ) = plural form; / = alternative form

app.	appendix
art.	article
assn.	association
attrib.	attributed to
bull.	bulletin
ch. / chap. (chs. / chaps.)	chapter
col. (cols.)	column
cont. / contd.	continued
dept.	department
dev.	developed by
dir.	directed by, director
div.	division
doc. (docs)	document
ed.	edited by, editor, edition
eds.	editors, editions
enl.	enlarged
eq. (eqs.)	equation
ex.	example
fig. (figs.)	figure
govt.	government
illus.	illustrated by, illustrator, illustration
inc.	incorporated, including
inst.	institute
intl.	international
jour.	journal
ms. (mss.)	manuscript
natl.	national
No. (Nos.)	number
p., (pp.)	page
pl.	plate, plural

(continued)

23.3 abbreviations found in bibliographies (cont.)

proc.	proceedings
reg.	registered, regular
resp.	respectively
rev.	revised by, revision; review, reviewed by
rpt.	reprinted by, reprint
sched.	schedule
sec. / sect.	section
ser.	series
sess.	session
soc.	society
supp.	supplement
tab.	table
trans.	translated by, translator, translation
vers.	version
vol. (vols.)	volume

23.4 common Latin expressions and abbreviations

There are no standard rules on the usage of Latin words and abbreviations. Below are just some general guidelines:

- Unless frequently used in your discipline, avoid less common Latin terms such as *ceteris paribus, sine non qua, mutatis mutandis.*
- Some experts suggest that certain Latin words and abbreviations should be avoided, since many people are unaware of their true meaning. There is certainly confusion between *e.g.* and *i.e.* (13.10).
- To decide whether you need to italicize a Latin word, check with your journal's instructions to authors and / or look in papers published by that journal. I suggest not using italics with: *e.g., et al., etc., i.e., per, versus, vs.,* and *vice versa*.

LATIN EXPRESSION	EQUIVALENT IN ENGLISH USAGE
a fortiori	with a stronger reason; if one fact exists then a second fact is even more true
a posteriori	from what comes after, a conclusion based on induction
a priori	evident by logic alone on the basis of what is already known
ab initio	from the beginning
ad hoc	created for this particular purpose only
ad libitum	without any advanced preparation, at the discretion of the researcher
anno domini (AD)	in the year of our lord (indicates the Christian era)
ceteris paribus	other things being equal
c. / ca. / circa	around, approximately
confer (cf.)	compare
de facto	in fact, in reality
erratum / errata	mistake / mistakes
et altri (et al.)	and others, and co-workers
et cetera	etcetera, and so on
et sequens (et seq.)	and the following
ex ante	before the fact, beforehand
ex post (facto)	after the fact, afterwards
exempi gratia (e.g.)	for example, for instance, such as

(continued)

23.4 common Latin expressions and abbreviations (cont.)

LATIN EXPRESSION	EQUIVALENT IN ENGLISH USAGE
ibidem	in the same place
id est (i.e.)	that is, that is to say
idem (id.)	the same
in silico ('modern' Latin)	via computer simulation
in situ	in its original place
in vitro	taking place outside a living organism
in vivo	within a living organism
inter alia	among other things
ipso facto	by the fact itself
modus operandi	characteristic method of working
mutatis mutandis	the necessary things having been changed i.e. this proof applies in more general cases
nota bene (NB)	NB, note that
per annum (p.a.)	for each year
per capita	per head
per diem (p.d.)	by the day
per impossibile	a proposition that cannot be true
per se	intrinsically, in itself
post mortem	autopsy
prima facie	on its face, i.e. a conclusion drawn only from the appearance of things
pro rata	proportionally
pro tempore	for the time being
quod et demonstrandum (QED)	that which was to be demonstrated
(reductio) ad absurdum	reduction to absurdity (disproof of a proposition by showing that it leads to an absurd conclusion)
sensu lato	in its broadest sense
sine non qua	essential condition
verbatim	without any changes to the original wording
versus	versus, vs., against
via	through, by means of
vice versa	vice versa, the other way round
videlicet (viz.)	viz, namely

24 Capitalization: headings, dates, figures etc.

24.1 titles and section headings

Both for titles and section headings, your choice will depend on your journal's style.

1. Use capital letters (upper case letters) for all words in the main title of a document except for the words below, unless they are the first word:
    - *a* and *the*
    - *it*
    - *and*
    - all prepositions (*by, from, of* etc.)
2. Alternatively, just use upper case for the first letter of the first word, and the rest in lower case. Section headings tend to follow this format.
3. Do not use a period (.) at the end of a title.

	YES	NO
1	A **G**uide to the **U**se of **E**nglish in **S**cientific **D**ocuments	A Guide **T**o **T**he Use Of English **I**n Scientific Documents
2,3	A guide to the use of English in scientific documents	A guide to the use of English in scientific documents**.**

A. Wallwork, *English for Academic Research: Grammar, Usage and Style*,
DOI 10.1007/978-1-4614-1593-0_24, © Springer Science+Business Media New York 2013

24.2 days, months, countries, nationalities, natural languages

1. Days, months, countries, nationalities and languages all have an initial capital letter.
2. Be careful with the use of *north(ern), south(ern), east(ern)* and *west(ern).* These only require initial capitalization when these are official regions shown on a map or atlas. For example, North Korea and South Korea are two separate nations.
3. When you want to refer to a geographical area of a country, then you can use two forms, e.g. *southern France, the south of France.* Note that in both cases there is no initial capitalization.
4. The following generally have initial capitalization: *the West, the Middle East, the Far East*. You can write either *the northern hemisphere* or *the Northern Hemisphere* (likewise with *southern*).

	YES	NO
1	The new versions in **English** and **Arabic** will be released on **Monday**, 10 **March** throughout **Egypt** and **Saudi Arabia**.	The new versions in **english** and **arabic** will be released on **monday**, 10 **march** throughout **egypt** and **saudi arabia**.
2	This species is found in **North Korea**, **East Timor**, and some parts of **South America**.	This species is found in **South Japan**, **East India** and some parts of **south America**.
2	This species tends to be found in the **north** and **west** of the island.	This species tends to be found in the **North** and **West** of the island.
2	The languages spoken in **northern Turkey** are quite disparate.	The languages spoken in **Northern Turkey** are quite disparate.
3	I love it when conferences are located in the **south** of **France**.	I love it when conferences are located in the **South** of **France**.
4	Uugter [67] reveals the total lack of morality in the **West** and compares it to the **Far East** where …	Uugter [67] reveals the total lack of morality in the **west** and compares it to the **far east** where …

24.3 academic titles, degrees, subjects (of study), departments, institutes, faculties, universities

1. Titles of job positions generally have an initial capital letter, particularly in formal documents (e.g. CVs, biographies for conferences) and when the position is held only by one person (in such cases *a / an* is not required). If the position is held by more than one person (*a / an* required), then initial capitals are not necessary though they may still be found.
2. Titles of degrees that are followed by the subject of study have an initial capital letter.
3. Subjects (mathematics, anthropology, history) have no initial capitalization when they are being talked about as subjects of study. However, when they are part of the name of a department, institute or faculty, they require initial capitalization.
4. The terms *department, institute, faculty* and *university* (and similar words) only require capitalization when referring to a specific department, university etc. The rules of initial capitalization for each individual word in the name of a department follow the rules given in 24.1.1.

	INITIAL UPPER CASE LETTERS	INITIAL LOWER CASE LETTERS
1,3	She is now **A**ssociate **P**rofessor at Nanjing University of Traditional Chinese Medicine.	He is an **a**ssociate **p**rofessor at Nanjing University of Traditional Chinese Medicine.
2	Short resume: Professor Wang has a **B**achelor of **A**rts in medicine, and a **M**aster's in alternative medicine.	I think she's got a **b**achelor's and a **m**aster's.
3	From 1891 to 1931 he was Professor of **M**athematics and **D**escriptive **G**eometry at the Technical University of Munich.	He studied **m**athematics and **i**nformation **e**ngineering before doing his Ph.D.
4	The **D**epartment of **S**ociology offers the following courses:	Our **d**epartment offers the following courses:
4	The **F**aculty of **E**conomics at the University of Bangkok has a long history of ...	Courses typically offered by **e**conomics **f**aculties and **e**ngineering **f**aculties include:

24.4 *figure, table, section* etc.; *step, phase, stage* etc.

1. When you refer to numbered sections, figures, tables, appendices, schedules, clauses, steps, phases, stages etc., capitalize the initial letter.
2. Do not capitalize the initial letter of *section, figure, table, appendix, schedule, clause, stage* etc. when there is no number associated.

Not all journals adopt the policies indicated in Rules 1 and 2 above.

	YES	NOT RECOMMENDED
1	See **Section** 2 for further details.	See the **section** 2 for further details.
1	See **Step 1** above.	See **step 1** above.
2	See the **appendix** for further details.	See the **Appendix** for further details.

24.5 keywords

In some documents, such as specifications and contracts, you may need to distinguish between different research units, users, projects, products etc. In such cases, initial capitalization is useful to make these keywords stand out for the reader.

CLEAR	LESS CLEAR
There are two types of user. Hereafter they will be referred to as **User A** and **User B**.	There are two types of user. Hereafter they will be referred to as **user a** and **user b**.
This will be the task of **Research Unit** 1.	This will be the task of **research unit** 1.
The two parties shall be referred to as the **Vendor** and the **Supplier**.	The two parties shall be referred to as the **vendor** and the **supplier**.
In the first phase, two prototypes will be developed: a prototype for automatically connecting to banks (hereafter, **Prototype** 1), and a prototype for risk management (**Prototype** 2).	In the first phase, two prototypes will be developed: a prototype for automatically connecting to banks (hereafter, **prototype** 1), and a prototype for risk management (**prototype** 2).

24.6 acronyms

All the letters of acronyms (22) have capital letters.

24.7 *euro, the internet*

Euro and *internet* are found both with (i.e. *Euro, the Internet*) and without initial capitalization.

25 Punctuation: apostrophes, colons, commas etc.

25.1 apostrophes (')

1. The main use of the apostrophe is to form the genitive (2). The only other use is if you want to make it clear to the reader how a word is constructed.
2. Do not use an apostrophe to make Acronyms and dates plural.
3. Contracted forms are not generally used in research manuscripts.

	YES	NOT RECOMMENDED
1	In my email I **cc'd** the co-authors who all have **PhD's**. cc'd = carbon copied	In my email I **ccd** the co-authors who all have **PhDs**.
1	A common mistake with the word 'aardvark' is to forget that it begins with two **A's**.	A common mistake with the word 'aardvark' is to forget that it begins with two **As**.
2	We bought six **PCs**.	We bought six **PC's**.
2	Our institute was founded in the 198**0s**.	Our institute was founded in the 198**0's**.
3	**Let us** now turn to Theorem 2, where we **will** learn that **it is** essential to …	**Let's** now turn to Theorem 2, where **we'll** learn that **it's** essential to …
3	The experiment **cannot / could not** be repeated.	The experiment **can't / couldn't** be repeated.

A. Wallwork, *English for Academic Research: Grammar, Usage and Style*,
DOI 10.1007/978-1-4614-1593-0_25, © Springer Science+Business Media New York 2013

25.2 colons (:)

1. The most common use of a colon in a research paper is to introduce a list.
2. Use a colon to divide up a two-part title of a paper or presentation. The word following the colon can either be capitalized or not. In such cases, a dash (25.5) could be used instead of a colon.
3. Avoid using a colon to add further thoughts or explanations if this will avoid creating an unnecessarily long sentence.
4. A colon may be useful to highlight a contrast, again providing this does not create a long sentence.

	YES	ALSO POSSIBLE
1	The following countries were involved in the **treaty: Turkey**, Armenia …	
2	Communicative language **teacher: The** state of the art Space **debris: the** need for new regulations	Ethical management in **banking – does** the presence of females make the difference?
3	This problem was first identified in the **1990s: in** the Sudan it was not noticed until 2013.	This problem was first identified in the **1990s. In** the Sudan it was not noticed until 2013 and in fact this led to serious problems with …
4	X can be used as an **identifier: Y** cannot.	X can be used as an **identifier. Y** cannot.

25.3 commas (,): usage

Commas should be used in the following situations:

1. To separate two dependent clauses. This is often the case with clauses introduced by *if, when, as soon as, after* etc.
2. To avoid initial confusion on the part of the reader. For example, in the NO example it initially seems that the water boils the specimen.
3. After sentences that begin with an adverb that is designed to attract the reader's attention (e.g. *clearly, interestingly*) or a link word that indicates you are adding further information or talking about a consequence (e.g. *consequently, in addition*).
4. In non-defining relative clauses (7.2.2).
5. If you have a list more than two items use a comma before *and* (13.4). The comma highlights that the penultimate and last element are separate items.

	YES	NO
1	When the specimen is **dry, remove** it from the recipient.	When the specimen is **dry remove** it from the recipient.
2	If the water **boils, the** specimen will be ruined.	If the water boils **the specimen** will be ruined.
3	**Surprisingly,** the results were not in agreement with any of the hypotheses. **Moreover,** in many cases they were the exact opposite of what had been expected.	**Surprisingly** the results were not in agreement with any of the hypotheses. **Moreover** in many cases they were the exact opposite of what had been expected.
4	The **Thames,** which runs through **London,** is England's longest river.	The Thames **which runs through London** is England's longest river.
5	There are three advantages of this: costs are lower, deadlines and other constraints are more easily **met, and** customers are generally happier.	There are three advantages of this: costs are lower, deadlines are more easily **met and** customers are generally happier.

25.4 commas (,): non usage

Commas should be avoided or limited if the sentence contains:

1. Twenty words or more. Consider rearranging the sentence or writing two separate sentences.
2. A series of very short phrases all separated by commas. Consider rearranging the sentence into longer phrases with fewer commas.
3. A long a list of items, which itself contains subgroups. It is better to use semicolons (25.11.12) to separate the various subgroups.
4. A series of nouns and the first and second noun are not related. Instead, begin a new sentence after the first noun, otherwise the reader will think that the nouns are all part of the same series.
5. In a defining relative clause (7.2.1).

	YES	NO
1	If the iodine solution does not turn to this color when added to a **food, this** indicates that starch is not present in the food.	**If, when** the iodine solution is added to **food, it** does not turn this **color, this** indicates that starch is not present in the food.
1	This application was developed specifically for this purpose. It can be used on most platform**s,** **f**or example XTC and B4M**E. It** can also be used with …	This application**, which** was developed specifically for this purpose**, can** be used on most platform**s f**or example XTC and B4M**E, it** can also be used with …
2	If Y is installed before **X, this** may cause damage.	Damage may be caused if X is installed **after, rather** than **before, the** installation of Y.
		If Y is installed **before, rather** than **after, installing Y, then** this may cause damage.
3	We used various sets of characters: A, B and **C; D**, E and **F; and** X, Y and Z.	We used various sets of characters: A, B and C, D, E and F and X, Y and Z.
4	Each row in the page represents an individual **record. The** information and the features provided enable the user to control, monitor and edit the records created.	Each row in the page represents an individual **record, the** information and the features provided enable the user to control, monitor and edit the records created.
5	The **student that** gets the top marks is awarded the prize.	The **student, that** gets the top marks is awarded the prize.

25.5 dashes (_)

Use a dash:

1. To avoid excess use of commas or brackets / parentheses in explanations. It is stronger than two commas, but lighter than parentheses. However it would usually be better to split up the sentence into smaller parts.
2. For afterthoughts to a final comment.

	YES	BETTER ALTERNATIVE
1	Taking this process into account, we would expect undesirable products – that is, unneeded doses (large pairs of isomers) – to form in the donor atoms.	Taking this process into account, we would expect undesirable products to form in the donor atoms. These products consist of unneeded doses, i.e. large pairs of isomers.
2	X does not, in fact, correspond to Y – and this is what we had suspected.	X does not correspond to Y. In fact, this is what we had suspected.
		X does not correspond to Y, thus confirming our suspicions.

25.6 hyphens (-): part 1

Use a hyphen when you:

1. Join two nouns together to form an adjective to describe another noun. Note: do not use a plural *s* on the noun that is acting as an adjective.
2. Use a word that acts as a prefix to the following word.
3. Have a series of prefixes referring to the same noun.
4. Prefix a word with *non* – this is not a rule and not all authors follow it.
5. Prefix a capitalized noun.
6. Refer to mixtures and analyses that combine two elements.

	YES	NO
1	A 30-**year**-old patient with one **six-fingered** hand.	A **30 years old** patient with one **six-fingers** hand.
2	To avoid **time-consuming** decisions, we used **row-based** flashing.	To avoid **time consuming** decisions, we used **row based** flashing.
3	Control of the interaction is **user-** not **application-**driven.	Control of the interaction is **user** not **application** driven.
4	These are **non-essential** items.	
	These are **non essential** items.	
	These are **nonessential** items.	
5	They made an assessment of soil depletion in **sub-Saharan** Africa.	They made an assessment of soil depletion in **sub Saharan** Africa.
6	We used **chemical-physical** analyses to determine the relative values in the **hydrogen-oxygen** mixture.	We used **chemical physical** analyses to determine the relative values in the **hydrogen oxygen** mixture.

25.7 hyphens (-): part 2

Use a hyphen when you:

1. Join a noun to a preposition (*clean-up, back-up*), but do not to join a verb to a preposition (*to clean up, to back up*).
2. Need to clarify any ambiguity.
3. Note: nouns, adjectives and prepositions only need to be joined together by a hyphen when in combination they act as adjectives that describe the following noun. If there is no following noun, then no hyphens are required (though this rule is frequently ignored).

	YES	NO
1	When the machine is **started up**, make sure …	When the machine is **started-up**, make sure …
	This feature is only available at **start-up**.	
2	This is a little **used-car**.	This is a **little used car**.
	A second-hand car that is small in size.	
	This is a **little-used** car.	
	A car that has been rarely driven.	
3	We present three **state-of-the-art solutions** to this **well-known** problem.	We present three **state of the art solutions** to this **well known** problem.
3	Automatic translation: the **state of the art**	Automatic translation: **the state-of-the art**

25.8 parentheses ( )

When readers see a phrase in parentheses, they may assume that the information contained therein is not very important. Don't use parentheses when it would be less distracting for the reader if you used a separate phrase. Use parentheses:

1. With acronyms and abbreviations. Put the full form outside the parentheses, and the acronym inside.
2. To give examples in the form of short lists, when this list appears in the middle of the phrase.
3. If a parenthesis appears at the end of a sentence, the period (.) should come after the parenthesis.

For more details on the types and usage of brackets see: http://en.wikipedia.org/wiki/Bracket.

	YES	NO
1	This is based on a **first in first out (FIFO)** policy.	This is based on a **FIFO (first in first out)** policy.
2	This is only true of three **countries (**i.e. Libya, Syria and Jordon**) and** for the purposes our study can be ignored.	This is only true of three **countries i.e.** Libya, Syria and **Jordon and** for the purposes our study can be ignored.
3	If there is no following noun, then no hyphens are required (though this rule is frequently **ignored).**	If there is no following noun, then no hyphens are required (though this rule is frequently **ignored.)**

25.9 periods (.)

1. Periods are not normally used at the end of titles or headings.
2. Periods are used in captions after the words Figure, Table etc., and in the related captions themselves.
3. If a word like *etc.* appears at the end of a sentence it only requires one period.
4. A set of three (or more) periods can be used to indicate that the preceding items are just some examples and there may be others. Using *e.g.* and *etc.* as well as the three dots is not necessary.

	YES	NO
1	A model for assessing the level of complexity in a **manuscript**	A model for assessing the level of complexity in a **manuscript.**
1	Materials and **Methods**	Materials and **Methods.**
2	Figure **1.** Transgene **structure.** Schematic representation of the fragment microinjected into the **nuclei.**	Figure **1** Transgene **structure** Schematic representation of the fragment microinjected into the **nuclei**
3	Various grammatical points are covered: tenses, adjectives, agreement **etc.**	Various grammatical points are covered: tenses, adjectives, agreement **etc..**
4	Various languages can be used **(C++, Java, …)** on most types of hardware **(IBM, Apple, …)**.	Various languages can be used **(e.g. C++, Java, …)** on most types of hardware **(IBM, Apple, … etc).**
	= Various languages can be used **(e.g. C++ and Java)**, and most types of hardware **(e.g. IBM and Apple).**	

25.10 quotation marks (‘ ’)

Check to see if your journal uses single (‘....’) or double (“....”) quotation marks. The rules for directly quoting the work of other authors vary from discipline to disciple and journal to journal

Below are just some examples.

1. If the quotation is short, incorporate it into the main text.
2. If the quotation is long, begin a new paragraph and indent the paragraph.
3. Another use of quotation marks is to enclose words and phrases that you have used in a special way. Use single quotations in such cases.

	YES	ALTERNATIVE
1	Wallwork states “A maximum of 20 words should be used per sentence” (Wallwork 2014). This implies that …	According to Wallwork (2014) “A maximum of 20 words should be used per sentence.” This implies that …
1	To determine “the best way to respond to referees without aggravating them” (Wallwork 2015) we devised a study based on a database of 476 replies to referees reports.	In order to determine what Wallwork (2015) posits as “the best way to respond to referees without aggravating them”, we devised a study based on a database of 476 replies to referees reports.
2	In her seminal work, Southern begins by saying: blah blah blah blah blah blah blah blah blah blah blah blah blah blah blah blah …	
3	We call this phenomenon ‘venting’, which is a variation of the so-called ‘wind synergism’.	

25.11 semicolons (;)

1. Use semicolons in lists that contain a series of phrases.
2. Use semicolons to make it clear which elements belong together in a series of lists.
3. Use semicolons to create a longer pause in the reader's assimilation of the sentence. This device should only be used rarely, given that it is likely to lead to the creation of a long sentence.
4. Do not join two independent clauses with a semicolon. Instead, make two simple, separate sentences.

	YES	AVOID
1	Substances are transported in living organisms as: (1) solutions of soluble **nutrients;** (2) solids in the form of food **particles;** (3) gases such as …	Substances are transported in living organisms as: (1) solutions of soluble **nutrients,** (2) solids in the form of food **particles,** (3) gases such as …
1	Figure 1. Three types of classroom arrangements: *a*, **traditional;** *b*, **circle;** *c*, U-shaped.	Figure 1. Three types of classroom arrangements: *a*, **traditional,** *b*, **circle,** *c*, U-shaped.
2	Several countries are participating in the project, in the following groups: Spain, Cuba and **Argentina; France**, Morocco and Senegal; and the Netherlands and Indonesia.	Several countries are participating in the project, in the following groups: Spain, Cuba and **Argentina, France**, Morocco and Senegal, and the Netherlands and Indonesia.
3	Sensory inputs merely modulate that **experiment; they** do not directly cause it.	
	Used here to create contrast between 'modulate' and 'cause'.	
3	The pitfalls described in this article have been known for many **years; our** work attempts to solve them.	
	Connects previous knowledge of Xs with author's own explanation of them.	
4	Users can search the entire **database. There is also a** special alert mechanism to inform administrators …	Users can search the entire **database; a** special alert mechanism is also provided that informs that administrator …

25.12 bullets: round, numbered, ticked

Bullets are rarely found in research papers. This is unfortunate as their use would often facilitate reading

1. Use round bullets when the sequence of the items is not important.
2. Use numbered bullets when the sequence of the items is important and to describe procedures.
3. Ticked bullets are sometimes used in reports and presentations to list what actions have already been taken in, for example, a project.

	YES	NO
1	To install the system you need: • Version 5.6 of Technophobe • Version 1.2 of Monstermac • Version 9.7 of SysManiac	To install the system you need: 1. Version 5.6 of Technophobe 2. Version 1.2 of Monstermac 3. Version 9.7 of SysManiac
2	The project is organized into three phases: 1. Specifications 2. Design and development 3. Release	The project is organized into three phases: • Specifications • Design and development • Release
3	We have made the following changes: √ two new tables added √ figures renumbered √ Appendix 2 removed	Conclusions We believe our approach has three major advantages: √ low cost √ easily adaptable √ quick set up times

25.13 bullets: consistency and avoiding redundancy

Within the same list or set of bullets:

1. Always begin with the same grammatical form. Use an introductory phrase that can always be followed by the same grammatical type (preferably an infinitive or a gerund).
2. Use the same style of punctuation and capitalization (there are no standards for this).
3. Avoid repeating unnecessary words.

	YES	NO
1	This would involve the following: • **acquiring** information on … • **understanding** the importance of … • **highlighting** any deficiencies in … All the bullets are -ing forms.	This would involve the following: • **the acquisition** of information on … • **understanding** the importance of … • **an ability to highlight** any deficiencies in …
1	These data are used to: • **acquire** information on … • **understand** the importance of … • **highlight** any deficiencies in … All the bullets are in the infinitive form.	These data are used: • **for the acquisition** of information on … • **to understand** the importance of … • **for highlighting** any deficiencies in …
2	There are three ways to learn English: – **find** a good teacher – **buy** DVDs and learn at home – **marry** a native English speaking person The first word of each bullet is in lower case.	There are three ways to learn English: – **find** a good teacher, – **Buy** DVDs and learn at home; – **Marry** a native English speaking person
3	… the decomposition into individual **modules**: • Settings Input • Platform Input • Engine	… the decomposition into individual **modules**: • Settings Input **module** • Platform Input **module** • Engine **module**

26 Referring to the literature

26.1 most common styles

There are four main ways to refer to other authors.

1. Begin the phrase with the author's name. This is the easiest style to use for authors and the most readable. This style is also useful for comparing authors.
2. Begin with the reference and then immediately give the name of the author. This is similar to the first style, and is particularly useful when you are referring to more than one work by the same author.
3. End with the author and / or the reference. Sometimes this can be a heavy construction because it involves the use of the passive. However it is useful when several papers are being referred to.
4. The author is not mentioned, only the reference number. This is potentially ambiguous, see 26.2.

To learn how to write the bibliography, see your journal's style guide or instructions to the author.

	EXAMPLES
1	Wallwork [2012] stated x = y.
	Huang [2013] agrees with this statement, but Xanadu [2014] does not.
2	In [6] Wallwork stated that x = y. Then in [9] he added that x + 1 = y + 1.
3	A proposal for a conference on this topic was put forward by Tang [2014].
3	This is not the first time that such a proposal has been put forward [Himmler, 2012; Goldberg, 2013].
	This is not the first time such a proposal has been put forward [6, 27, 33].
4	This proposal was first put forward in [6].
	In [6] a proposal for a conference on this topic was put forward.

DOI 10.1007/978-1-4614-1593-0_26, © Springer Science+Business Media New York 2013

26.2 common dangers

1. Only using a reference without mentioning the author's name is certainly the most concise form. But it has a major disadvantage. It forces the reader to break off reading the text and go to the bibliography to see which author is being referred to. The reader may also need to ascertain whether the author is referring to his / her own work. It can also be ambiguous when the phrase *the author/s* is used – is this a reference to the author/s of the current paper or to other authors in the literature?
2. If you refer to a previous paper that you have written, make sure it is clear that this paper was written by you and not by another author. Just mentioning your name is not enough because the reader may have forgotten that the paper that he/she is reading is by you.
3. Within the same manuscript do not mix the year format with the bibliographical number format. Your choice will depend on the journal.

Note: the best solution may be to use more than one form in order to create variety. A review of the literature can be very tedious if every sentence either begins or ends with the name of an author or a reference. Also, you may need to change the focus from author to findings and vice versa.

	YES	AMBIGUOUS OR WRONG
1	In [6] Wallwork put forward a proposal for the scientific community to allow personal forms. = another author In [6] we put forward a proposal for the scientific community to allow personal forms. = the author of the current paper	In [6] the author put forward a proposal for the scientific community to allow personal forms.
2	In a previous paper [Gomez, 2], we found that x = y.	In [Gomez, 2], it was found that x = y.
3	In [6] Wallwork stated that all journals should allow the use of personal forms. Two years later he added that the ISO should set some standards regarding the style of bibliographies [9].	In [6] Wallwork stated that all journals should allow the use of personal forms. Two years later he added that the ISO should set some standards for scientific writing [Wallwork, 2014].

26.3 punctuation: commas and semicolons

Below are some examples of how to punctuate references within the main text. They should be considered as being typical usages rather than rules.

SUGGESTED USAGE	EXAMPLES
one author: name + comma + year	Wallwork, 2015
two authors: name1 'and' name2 + year	Wallwork and Southern, 2016
three authors: name1 + comma + name2 'and' name3 + year (Note: writing the names of three authors is quite unusual)	Wallwork, Brogdon and Southern, 2016
three or more authors: name1 + et al.	Wallwork et al., 2016
two or more references: ref1 + semicolon + ref2 + semicolon etc.	Wallwork et al., 2016; Sanchez, 2017; Poplova, Huang and Sun, 2018
several works by same author: name + comma + year1 + comma + year2 etc.	Wallwork, 2012, 2014, 2016

26.4 punctuation: parentheses

Some journals use (rounded parentheses), others use [square parentheses].

SUGGESTED USAGE	EXAMPLES
when the author is the subject of the verb: name + year in parentheses. Alternatively: name + reference number in parentheses	Wallwork [2012] suggests that … Wallwork [6] suggests that …
when the author is not the subject of the verb: both name and year in parentheses	It has been suggested that one plus two is equal to four (Moron, 2011).

26.5 *et al*

1. Most journals use *et al* if there are three or more authors. An alternative to *et al* is *co-workers* or *coworkers.*

2. *Et al* is put in italics in many journals. *et al* is followed by a period in some journals.

	YES	ALSO POSSIBLE
1	Wallwork et al [2016] put forward a proposal for the scientific community to allow personal forms.	Wallwork and co-workers [2016] put forward …
2	Wallwork et al [2016] suggested that …	Wallwork *et al* [2016] suggested that …
		Wallwork *et al.* [2016] suggested that …

27 Figures and tables: making reference, writing captions and legends

27.1 figures, tables

1. Use a capital F and T for *Figure* and *Table* when these are associated with a number, use lower case when they are not associated with a number.
2. The abbreviation for *Figure* and *Figures* is *Fig.* and *Figs.*, avoid abbreviating the word *Table*.
3. Be concise when introducing or making reference to a figure or table.
4. Where possible use the active form rather than the passive.
5. Use *as* not *as it* (13.5).

	YES	NOT RECOMMENDED (1–4), NO (5)
1	See Figure 1 and Table 2.	See figure 1 and table 2.
2	See Fig. 1a and Figs. 2a and 2b.	See Fig. 2a and "b.
3	Figure 2 below **shows** the initial settings.	The following figure (Figure 2) **gives a schematic overview of** the initial settings.
3	Figure 3 **shows** the architecture.	The **snapshot depicted in** Figure 3 **shows a view of** the architecture.
3	For **details, see** [Kyun, 2013].	For **further details on this topic, the reader is kindly invited to refer to** [Kyun, 2013].
4	Figure 2 below **shows** the initial settings.	The initial settings **are shown** in Figure 2 below.
		In Figure 2 the initial settings **are shown**.
5	**As** can be seen in the figure below …	**As it** can be seen in the figure below …

A. Wallwork, *English for Academic Research: Grammar, Usage and Style*,

DOI 10.1007/978-1-4614-1593-0_27,

27.2 legends

There are no standard rules for writing legends to figures and tables. I recommend making your caption grammatically correct, for example use articles and prepositions where required. Below is a possible format:

Figure 1. The main characteristics of the shock absorbers.

That is to say:

- Initial capital letter for *figure, table, appendix* etc. Do not use an abbreviation.
- After the number put a full stop.
- Initial capital letter for the first word in the description.
- End the line with a full stop.

27.3 referring to other parts of the manuscript

1. When you refer to something within the same document, avoid expressions such as *above, below, later, on the previous page, in the next section*. Refer the reader to a specific heading and page number.
2. When you want to refer back to something you wrote about in the previous paragraph, use: *noun + mentioned above* or *above-mentioned + noun* (note the use of the hyphen 25.6).
3. H*ereafter* is a useful word when you have a long term that you want to abbreviate, and this abbreviation will then be used in the rest of the document.
4. T*he following* is followed by a noun.

	YES	NOT RECOMMENDED (1), WRONG (2–5)
1	As mentioned in **Section 2**, this procedure is …	As mentioned **above**, this procedure is …
1	This procedure is extremely complex and is described in **Section 4**.	This procedure is extremely complex and is described **later**.
2	The function **mentioned above** is …	The function **above mentioned** is …
	The **above-mentioned** function is …	
3	This feature is known as an 'automatic rendering and masking agent' **hereafter** ARM agent.	This feature is known as an 'automatic rendering and masking agent' **in the following** ARM agent.
4	The **following versions** can be used:	The versions that can be used are **the following**:
	The versions that can be used are **as follows**:	

28 Spelling: rules, US versus GB, typical typos

28.1 rules

RULE	ROOT WORD	PRESENT / PLURAL	PAST PART. / COMPARATIVE	-ING FORM	-LY / -ABLE
1 S: 1 V + 2 C	work	works	worked	working	workable
	quick		*quicker*		quickly
1 S: 2 V + 1 C	heat	heats	heated	heating	heatable
	great		*greater*		greatly
1 S: 1 V + 1 C	stop	stops	sto**pp**ed	sto**pp**ing	unsto**pp**able
	glad		*gla**dd**er*		gladly
2 S: stress on S1	cover	covers	covered	covering	coverable
2 S: stress on S2	prefer	prefers	prefe**rr**ed	prefe**rr**ing	preferable
-ch	reach	reaches	reached	reaching	reachable
C + e	note	notes	noted	no**ting**	no**tab**le
	enlarge	enlarges	enlarged	enlar**ging**	enlargeable
			larger		largely
-c + e	replace	replaces	replaced	repla**cing**	replaceable
-e + e	agree	agrees	agreed	agreeing	agreeable
-i + e	tie	ties	tied	**ty**ing	untieable
-is	thesis	theses			
-l + e	sample	samples	sampled	sam**pl**ing	samplable
	simple		*simpler*		sim**ply**
-u + e	argue	argues	argued	arguing	arguable
-ic	panic	panics	pan**ick**ed	pan**ick**ing	trag**icall**y
-lic	public				publ**icly**

(continued)

A. Wallwork, *English for Academic Research: Grammar, Usage and Style*,
DOI 10.1007/978-1-4614-1593-0_28, © Springer Science+Business Media New York 2013

28.1 rules (cont.)

RULE	ROOT WORD	PRESENT / PLURAL	PAST PART. / COMPARATIVE	-ING FORM	-LY / -ABLE
-l	travel	travels	trave**ll**ed (GB)	trave**ll**ing (GB)	
			traveled (US)	traveling (US)	
	hill		*hillier*		hilly
-no	piano	pianos			
-o	forego	foreg**oes**		foregoing	foregoable
	potato	potat**oes**			
-sh	push	pushes	pushed	pushing	pushable
			pushier		
-ss	pass	passes	passed	passing	passable
V. + w	narrow		*narrower*		narrowly
-x	fix	fix**es**	fixed	fixing	fixable
C + y (1 S)	shy	sh**ies**	sh**ie**d	shying	shyly
			*sh**ie**r*		
C + y (2 S)	happy		*hap**pie**r*		hap**pil**y
	marry	mar**ries**	mar**rie**d	marrying	mar**ria**ble
V + y	enjoy	enjoys	enjoyed	enjoying	enjoyable
-zz	jazz	jazz**es**	jazzed	jazzing	jazzable
			jazzier		

Legend: S = syllable (2 S = two syllables), C = consonant, V = vowel

Notes:

- In the table above there are some words that are not in common use (e.g. *gladder, jazzable*), but are simply designed to highlight a spelling rule.
- Check with your journal whether American or British spelling is required. Ensure your spelling is consistently British or American. For more details see: http://en.wikipedia.org/wiki/Spelling_differences#Simplification_of_ae_and_oe.
- Some words can be spelled two ways: *ageing / aging, spelled / spelt, dreamed / dreamt, focussed / focused, focussing / focusing.*
- Some words have a different spelling depending on whether they are a noun or verb. Here are some typical examples: *imbalance* (n), *unbalance* (v); *practice* (n, GB English), *practise* (v, GB; v + n US).
- Many people often use *-ize* and *-ise* indifferently.

28.2 some differences in British (GB) and American (US) spelling, by type

GB	US	GB	US
-ae-, -oe-	-e-	anaemia, archaeology, anaesthesia	anemia, archeology, anesthesia
-amme	-am	programme	program
arte-	arti-	artefact (but: artificial, artist etc)	artifact
-ce	-se	defence, offence	defense, offense
-ce	-se	practice, licence (n). practise, license (v)	practise (n, v), license (n, v)
-edge-	-edg-	acknowledgements	acknowledgements
-elled	-eled	modelled, travelled	modeled, traveled
-ey	-ay	grey	gray
-ise, -yse	-ize, -yze	analyse, materialise, realise	analyze, materialize, realize
-ium	-um	aluminium	aluminum
-l	-ll	marvellous	marvelous
-ller	-ler	modeller, traveller	modeler, traveler
-oe	-e	oedema	edema
-ogue	-ge	analogue, catalogue, dialogue	analog, catalog, dialog
-our	-or	behaviour, colour, flavour	behavior, color, flavor
-ph-	-f-	sulphur	sulfur
-que	-k	cheque (money)	check
-re	-er	centre, fibre, metre	center, fiber, meter
-wards	-ward	backwards, forwards, towards	backward, forward, toward

28.3 some differences in British (GB) and American (US) spelling, alphabetically

GB	US
acknowledgements	acknowledgements
aluminium	aluminum
anaemia	anemia
anaesthesia	anesthesia
analogue,	analog
analyse	analyze
archaeology	archeology
artefact	artifact
backwards	backward
behaviour	behavior
catalogue	catalog
centre	center
cheque	check
colour	color
defence	defense
dialogue	dialog
empower	impower / empower
ensure	insure / ensure
fibre	fiber
flavour	flavor
forwards	forward
grey	gray
labelled	labeled
licence (n)	license (n, v)
license (v)	license (n, v)
marvellous	marvelous
materialise	materialize
metre	meter
modelled	modeled
modeller	modeler

(continued)

(continued)

GB	US
oedema	edema
offence	offense
practice (n)	practise (n, v)
programme	program
realise	realize
sulphur	sulfur
towards	toward
travelled	traveled
traveller	traveler

28.4 misspellings that spell-checking software does not find

Some misspellings will not be highlighted because they are words that really exist. When you have finished your document, do a 'find' and check if you have made any of the mistakes listed below. Note that these are just examples, there are many other possible mistakes of this type.

WORD	EXAMPLE	WORD	EXAMPLE
addition (n)	The addition of gold led to higher values.	addiction (n)	Their addiction to cannabis had let to behavioral problems.
analyzes / ses (v)	The software analyzes the data.	analyses (n pl., sing. *analysis*)	We carried out two analyses.
assess (v)	We assess the pros and cons.	asses (n pl)	Horses and asses (*equus asinus*).
context (n)	The meaning of a word may depend on the context.	contest (n)	This is basically a contest between males and females.
chose (inf. *choose*)	In the past we always chose this method because …	choice (n)	The rationale behind our choice was …
drawn (inf. *draw*)	Conclusions are drawn in Sect. 5	drown (inf)	The fish drown in the nets.
fell (inf. *fall*)	The tree fell on the house.	felt (inf. *feel*)	The patients said they all felt anxious.
filed (inf. *file*)	It is filed under 'docs'.	field (n)	The field of ICT is ever growing.
form (v)	We would like to form a new group.	from (prep)	Professor Yang comes from China.
found (inf. *find*)	We found very high values in …	founded (inf. *found*)	IBM was founded in 1911.
lose (inf.)	Companies may lose a lot of money.	loose (adj)	There is only a loose connection between the two.
rely (v)	We rely on CEOs to make good decisions.	relay (v, n)	This relays the information to the train's onboard computer
than (conj, adv)	This is better than that.	then (adv)	After Stage 1, we then added the liquid.
thanks (n pl)	Thanks are due to the following people:	tanks (n)	The fish were stored in water tanks.

(continued)

(continued)

WORD	EXAMPLE	WORD	EXAMPLE
though (adv, conj)	The overheads are high, though the performance is excellent.	tough (adj)	This is a tough question to answer.
through (prep)	This was achieved through a comparative study of …	trough (n)	Pigs eat from a trough.
two	Two replications were made.	tow (v)	The car is equipped to tow a caravan.
three (n)	Tests were repeated three times.	tree	Tests were conducted on an apple tree.
use (v, n)	We use a method developed by …	sue (v)	Patients frequently sue their physicians for malpractice.
weighed (inf. *weigh*)	The samples were dried and then weighed.	weighted (adj)	The weighted values were obtained by dividing the integral of the …
which (pronoun)	This worked well, which was surprising considering that …	witch (n)	Life often ended early for a witch in medieval times – burnt on the stake.
with (conj)	We worked with them in 2013.	whit (n)	Whit is a religious festival.

Erratum

English for Academic Research: Grammar, Usage and Style

Adrian Wallwork

A. Wallwork, *English for Academic Research: Grammar, Usage and Style*,
DOI 10.1007/978-1-4614-1593-0, © Springer Science+Business Media New York 2013

DOI 10.1007/978-1-4614-1593-0_29

The publisher regrets that in the original publication the word "Academic" was mistakenly left out of the book title in the front matter and the chapter opener of the print and online versions. "English for Research: Grammar, Usage and Style" is incorrect.

The correct title should read: "English for Academic Research: Grammar, Usage and Style"

This title has since been updated accordingly.

The updated original online version of this book can be found at
http://dx.doi.org/10.1007/978-1-4614-1593-0

Appendix 1: verbs, nouns, adjectives + prepositions

This appendix lists the following:

- irregular verbs (only the most commonly used in academia)
- verb + infinitive, or verb + *-ing*
- verb + preposition
- noun + preposition
- adjective + preposition

Legend:

[ ] = the past form and past participle, if there is only word this means that the past form and the past participle are the same

+ inf = this verb takes the infinitive

+ ing = this verb takes the -ing form

+ inf/ing = this verb takes both forms, possibly with a difference in meaning

n = noun

v = verb

/ = both forms are possible, but probably with a change in meaning

, = the word that follows the comma precedes the main word (e.g. *addition to, in* = in addition to)

abide [abode] by
ability + inf
able + inf
above -
absence of, in the
accept + inf
accompanied by
accordance with, in
according to
account for
accustomed to
achieved by
acquaint with
act as
act upon
action of X on Y
adapt X to Y
add up to
add X to Y
addition of X to Y
addition to, in
adequate for
adhere to
adherence to
adjacent to
adjust X to Y
advance, in
advantage in
advantage of X over Y
adverse to

A. Wallwork, *English for Academic Research: Grammar, Usage and Style*,
DOI 10.1007/978-1-4614-1593-0, © Springer Science+Business Media New York 2013

advise X to do Y
affiliate to
agree + inf
agree with
aid X to do Y
aim + inf
aim to
aimed at
allocate X to Y
allow for
allow X to do Y
allowance for
ally with
alteration in
alternative to
amenable to
amount to
analogous to
answer (n) to
answer (v) -
answer X
apart from
appeal to
appear + inf
append X to Y
apply X to Y
appointment with
approach to
approach, in this
appropriate for
approve of
arise [arose, arisen] from
arranged in/into
arrive (at)
article on/about
ask for X
ask X to do Y
assign X to Y
assimilate X into Y
assist in
assist X to do Y
assist X with Y
associate X with Y
assume that
assumed + inf
assumption, on/under the
attach X to Y
attempt (n) to + inf, at + ing
attempt (v) + inf
attention on
attention, give [gave, given] to
attract X to Y
attracted to/by
attribute X to Y
average, on
avoid + ing
aware of
axis, on an
balance X with/against Y
based on
basis, on the
be [was, been]
bear [bore, born] in mind
bear [bore, born] out (by)
become [became, become]
begin [began, begun] + inf/ing
begin [began, begun] with
believe in
belong to
below -
benefit from
benefit of, a
bind [bound] X to Y
birth to, give [gave, given]
bite [bit, bitten]
blame X for Y
bleed [bled]
blow [blew, blown]
book on/about
book, in a
borrow from
bottom, at the
bound to
bounded by
break [broke, broken]
breed [bred]
bring [brought]
broadcast [-cast/-casted]
build [built] on
burn [burnt/burned]
burst [burst]
calculate for
call attention to
candidate for
capable of
capacity for
caption to the figures
careful + inf
carry out
cash, in
catch [caught]
cater for

cause X to do Y
cease + inf/ing
challenged with
chance, by
chances of
change in
change X into Y
change X with/for Y
characterized by
charged to
charged with
check whether
choice, by
choose [chose, chosen] between/from
claim + inf
close to
closed to
clue to
clustered in
coefficient on
coerce X to Y
coincide with
collaborate with
collide with
colored [with]
combination of X and Y
combine X with Y
come [came, come]
command X to do Y
comment on
commit X to Y
common to
common with, in

comparable to/with
compare X to/with Y
comparison of X and Y
compatible with
compel X to do Y
compensate for
compliance with, in
comply with
composed of
comprised of/in
conceive of
concentrate on
concern
concerned with/about
concerned, as far as X is
concession to
conclusion, in
concur with
condition, under/in a
confer with
confer X on Y
confidence in
confine X to Y
conflict with
conform to
confront (X and Y)
confusion with/between
congratulate X on Y
connect to/with
connection with, in
conscious of
consent + inf
consequence of, (as a)
consequent from

conservative over
consider -
consign X to Y
consist in + ing
consist of
consistent with
constraint on
consult (with)
contact in
contact with, in
contained in
contaminated with
contingent to
continue + inf/ing
contradistinction to
contrary to
contrary, on the
contrast to, in
contrast, by
contribute to
contrive + inf
control, in
control, out of
converge to / in
convert X into Y
convert X to Y
convertible into
convey X to Y
convince X to do Y
cooperate for a purpose
cooperate in work
cooperate with X
coordinate X with Y
cope with

correct X to Y
correlate X with Y
correspond to
correspond with
cost [cost] -
count on
counteract by
coupled with
credit for
cut [cut]
deal [dealt] with
debate about
decide + inf
decide for/against
decide on
decompose X into Y
decrease in
deduce X from Y
defend X from Y
deficiency of X in Y
defined as
defined by
definition, by
degenerate into
delay + ing
delay in
deliver X to Y
demand that X do Y
denote X by/with Y
depend on
dependence on
depending on
deposit on/onto
deprive X of Y
derive X from/by Y
designated by
designed by
detach X from Y
detail, enter into
detail, in
deter X from Y
detriment of, to the
detriment to, without
develop X into Y
deviate from
deviation in
devoid of
devote to
diagnose X as being Y
die of
differ from
difference from/between
difference in
different from
differentiate between
difficulty in
difficulty, with
direct X to do Y
disagree with
disassociate from
discourage X from doing Y
discuss X with Y
discussion, under
dispense with
displacement of
dissolved in
distinct from
distinguish between X and Y
distinguish X from Y
divide (up) X into Y
divide by
do [did, done]
dominate over
doubt whether
downstream of
draw [drew, drawn] attention to
draw [drew, drawn] on
drawback of/to
dream [dreamt/dreamed] about
drink [drank, drunk]
drive [drove, driven] by
dry in
due to
duty to
ease, with
effect of X on Y
effect, bring [brought) into
elevate X to Y
embark on
emitted by
emphasis on
empty of
enable X to do Y
encourage X to do Y
end, to this
endowed with
enquire into
enroll in
ensue from
entail + ing
enter [into]
entitled to

entrust X with Y
envisage + ing
equal to
equate to
equate X with Y
equilibrium, in
equip X for Y
equipped with
equivalent to
essential to
event of, in the
evidence from
evidence of/for
examination, under
except for
exception of, with the
excess of X in Y
excess of, in
exchange X for/with Y
exclude X from Y
exert X on Y
exertion, by
expect X to do Y
experience in
experiment with
expert on, an
explain X to Y
explanation, in
expose X to Y
exposure to
expressed by
expressed in
extend X to Y
extension of

extent, to an
external to
extreme, at
faced with
fact, in
fail + inf
fail in + noun + ing
fall [fell, fallen]
fall in, a
fault in/with
feasibility of
feature of
feed [fed] X into Y
feel [felt]
fight [fought]
fill in/out
fill with
find [found] (to)
fit in
fit with
fit X into Y
fluctuation in
fly [flew, flown]
focus (X) on Y
follow on from
forbid [-bade, -bidden] X to do Y
force X to do Y
forecast [-cast / -casted]
foresee [-saw, -seen]
forget [-got, -gotten]
form of, in the
formed by
formed on

free + inf
free X of/from Y
freeze [froze, frozen]
front of, in
full of
function of, as a
fundamental to
generate X from/by Y
get [got, got/gotten]
give [gave, given] rise to
give [gave, given] XY
go [went, gone]
gradation, in
graduated in
grant X to Y
grind [ground]
grounds of, on the
group X into Y
grow [grew, grown]
guarantee + inf
guarantee against
guarantee X that Y
guarantee X Y
guided by
hang [hung]
have [had]
hear [heard]
help X to do Y
help X with Y
hide [hid, hidden] X from Y
hit [hit]
hold [held] (true) for
hurt [hurt]
hypothesis, under a

identical to
immerse X into Y
immersed in
immunity to
impact on
impart X to Y
impermeable to
implicated in
imply + ing
importance to
impose X on Y
improve on
improvement in/on
incident upon
include X in Y
inclusive of
incompatible with
incongruous with
incorporate X into Y
increase in, an
increased by
indebted to
independent of
induce X to do Y
infected with
inferior to
influence (v) X
influence of X on Y
inherent in
initiate X into Y
inject X into Y
input into
input (inputted)
inscribe with
insert X into Y
insertion into
insight into
insist on
inspired by
instant, at an
instead of
integral with
intend + inf
intended for
interact with
interest in
interested in
interests of, in the
interfere with
internal to
interval, at
introduce in/into
introduce X to Y
invest (X) in Y
investigate (into)
investigation, under
invite X to do Y
involve + ing
involved in
irrespective of
isomorphic to
joined to
journal, in a
keep [kept]
key to
know [knew, known] of/about
lack of
last for
lay [laid] stress on
lead [led] X to do Y
lean [leant/leaned] on
learn [learnt/learned] + inf
least, at
leave [left]
left, on the
legend to the figures
lend [lent] force to
lend [lent] XY
let [let] X do Y
level, on a
liaise with
license X to do Y
light [lit]
light of, in the
likelihood of
likened to
limit X to Y
limit, within a
linear to
linked to
load X into/onto Y
look forward to
lose [lost]
loss of
made up of
magazine, in a
make [made] X do Y
manage + inf
map onto
map X on/onto Y
map, on a
match (v) -

maximum, at a
mean [meant] + inf
mean [meant] by
means of, by
measured in
mediate between
meet [met] (with)
middle, in the
minimum, at a
mislead [-led]
mistake [-took, -taken] X for Y
mistake, by
mix X with Y
modification to
modify X into Y
more than
most, at
motion, in
move X to Y
multiply by
nature, by
near -
necessity of
necessity, by
need + inf/ing
need for
neglectful of
neighbor of
next to
normal to
obey X
object to
oblige X to do Y
occasion, on an
occur in
offer to do X for Y
offer XY
open to
operation, in
opportunity + inf, for
opposed to, (as)
opposite -
opposite -
order of, in the
organize X into Y
originate from/by
orthogonal to
output (outputted)
overview of
owing to
painted [with]
par with, on (a)
parallel to/with
parallel, in
part of
participate in
partition X into Y
pattern, in a
pay [paid] attention to
pay [paid] X for Y
peculiar to
penetrate into
permeable to
permission to
permit X to do Y
perpendicular to
persist in
persistence in
persuade X to do Y
pertaining to
phone, on the
place of, in
plan + inf
play a part in
point of view, from
point out
point to (at)
point, at a
poor in
possession of, in
possibility of
power of, to the
practice, in
precedence over, have
precedence to, give
predicted by
predominate over
prefer X to Y
prefer x to y
preliminary to
preoccupied with
prepare X for Y
prepared + inf
prescribe X for Y
presence of, in the
preside over
press, at the
pressure, at a
pressure, under
pretext for
prevail over
prevent X from

principle, in
prior to
probability of
problem with
proceed + inf
proceed by + ing
proceed with
product of
profit from
progress, in
project X onto//upon Y
prompt X to do Y
proportion to, in
proportional to
propose + ing/inf
propose X to Y
protect X from/against Y
protective of/towards/against
protest against
prove [proved, proved/proven] X on Y
provide against
provide for
provide X with Y
provoke X to do Y
purpose, on
put [put] in/into
question, in
raise X by
raise X to
random, at
range, in the
rate, at a
rather than
ratio of X to Y
react to/with
read [read]
reason (n) why
reason for
recall + ing
recede from
recommend that X do Y
reduce X to
reduced to
refer X to Y
reference to, with
refine X into Y
regarded as
regardless of
regards, as
reinforce with
relate to
related to
relating to
relation to, in
relation with/between
relationship between/among
relative to
release X from Y
relief from
relief, in
relieve X from/of Y
rely on
remember + inf/ing
remind X to do Y
remove X from Y
reorganize X into Y
replace X by/with Y
reply to
report on/about
representative of
request (n) for
request X to do Y
require that X do Y
required for
research on/about/into
resemble -
resist + ing
resistance to
resistant to
respect (n) for
respect to, with
respect, in
respond to
response to, in
responsible for
responsive to
restrict X to Y
result from
result (v) in
result of, as a
review of/on
review, in a
rich in
ride [rode, ridden]
right, on the
ring [rang, rung]
rise [rose, risen]
rise in
risk + ing
risk of
risk to

role in, play a
room for
rule, as a
sake of, for the
same as
same time, at the
satisfied with
say [said] to
scale, on a
scope, beyond the
seal off/up
search for
see [saw, seen]
seeing as
select X from/by Y
send [sent] XY
sense, in a
sensitive to
separate X from Y
series, in
serve as
serve to
set [set, set]
shake [shook, shaken]
share X with Y
shares in
sharing of
shed [shed]
shield X from Y
shine [shone]
shoot [shot]
show [showed, shown] XY
shrink [shrank, shrunk]
shut [shut]
similar to
sit [sat, sat]
skilled in
slide [slid, slidden]
smell [smelt/smelled]
soluble in
solution to/of/for
solve X with Y
speak [spoke, spoken] to/with/about
specialist in
spell [spelt/spelled]
spend [spent] (time + ing)
spill [spilt/spilled]
spin [span, spun]
split [split] into
spoil [spoilt/spoiled]
sponsored by
spread [spread]
spring [sprang, sprung]
stand [stood] for
steal [stole, stolen]
step in
stick [stuck]
stimulate X to do Y
stop + inf (stop X in order to do Y)
stop + ing (stop X)
stop X from doing Y
stored in
stress on
strike [struck]
study on/of, a
study X
study, under
subject X to Y
subjected to
submit X to Y
subsequent to
substitute by/with/for
subtract x from y
succeed in
successful in
succession, in
suffer from
suggest doing X
suggest that X do Y
suitability of X for Y
suitable for
suited to
summary, in
superimposable to
superior to
supply X to Y
support for
survey of/on
susceptible to
swell [swelled, swollen]
swim [swam, swum]
switch from X to Y
sympathize with
synchronize X with Y
synchronous with
tailored for
take [took, taken] part in
take [took, taken] X from Y
take X into account
talk about
tally with

teach [taught] X to do Y
tear [tore, torn]
tell [told] XY
temperature, at
tend + inf
tendency to
tending to
terms of, in
tests on
thanks to
theory, in
think [thought] about / of
throw [threw, thrown]
tied to
together with
top, at the
trace out
transform X into Y
translate X into Y
transmit X to Y
transparent to
transverse to
trouble with
try + inf
turn X into Y
TV, on the
understand [-stood] how
undertake [-took, -taken] + inf
uniform in
unit of
unite X with/to Y
upstream of
urge X to do Y
vacuum, under
value, in
variance, at
variation in
vary in
vary with
verify whether
visualize + ing
vital to
vouch for
wait for X to do Y
want X to do Y
warn X about/against Y
watch X doing Y.
way + inf
wear [wore, worn]
whole, on the
wind [wound]
work on
worth + ing
write [wrote, written]
yield to

Appendix 2: Glosssary of terms used in this book

The definitions below are my definitions of how various terms are used in this book. They should not be considered as official definitions.

abbreviation	a shortened form of a word e.g. *info* rather than *information*
acronym	e.g. *url, www, NATO, IBM*
active (form)	use of a personal pronoun/subject before a verb, e.g. *we found that x = y* rather than *it was found that x = y*
adjective	a word that describes a noun (e.g. *significant, usual*)
adverb	a word that describes a verb or appears before an adjective (e.g. *significantly, usually*)
ambiguity	words and phrases that could be interpreted in more than one way
comparative	e.g. *better, happier, more intelligent*
conditional	e.g. *If I spoke perfect English, it would be easier to write papers*
countable noun	a noun that can be made plural by adding an *-s*, e.g. *books, students*
definite article	*the*
direct object	in the sentence *I have a book*, the book is the direct object
genitive	the possessive form of a noun, e.g. *Adrian's book*
gerund	the part of the verb that ends in – *ing* and that acts like a noun (e.g. *learning, analyzing*). The terms *gerund* and *-ing* form are used indifferently in this book
indefinite article	*a / an*
indirect object	in the sentence *I gave the book to Anna*, book is the direct object, and Anna is the indirect object
- *ing* form	the part of the verb that ends in – *ing* and that acts like a noun (e.g. *learning, analyzing*). The terms *gerund* and *-ing* form are used indifferently in this book
infinitive	the root part of the verb, e.g. *to learn, to analyze*)
link word, linker	words and expressions that connect phrases and sentences together (e.g. *and, moreover, although, despite the fact that*)
modal verb	verbs such as: *can, may, might, could, would, should*
noun	words such as: *a/the paper, a/the result, a/the sample.* If the noun can be made plural it is 'countable', if it only exists in the singular it is 'uncountable'

A. Wallwork, *English for Academic Research: Grammar, Usage and Style*,
DOI 10.1007/978-1-4614-1593-0,

paragraph	a series of one or more sentences, the last of which ends with a paragraph symbol (¶)
passive (form)	an impersonal way of using verbs, e.g. *it was found that* $x = y$ rather than *we found that* $x = y$
past participle	e.g. it was *found*, we have *found*, we have *seen*, they have *done*
phrasal verb	e.g. *back up, break down, look forward to, turn off, work out*
phrase	a series of words that make up part of a sentence
preposition	e.g. *to, at, in, by, from*
punctuation	.(period) , (comma); (semi-colon) : (colon) - (hyphen) () (brackets/ parentheses) ? (question mark) 'blah' (single quotes), "blah" (double quotes)
quantifier	e.g. *some, every, any, all, many*
relative pronoun	*who, which, that, whose*
sentence	a series of words ending with a period (.)
superlative	e.g. *best, happiest, most intelligent*
tense	future simple (*we will study, he will study* etc), present simple (*we study, he studies*), present continuous (*we are studying, he is studying*), present perfect (*we have studied, he has studied*), present perfect continuous (*we have been studying, he has been studying*), past simple (*we studied, he studied*), past perfect (*we had studied, he had studied*), past continuous (*we were studying, he was studying*)
uncountable noun	a noun that only exists in the singular, e.g. *information, feedback, software*
word order	the order in which nouns, verbs, adjectives and adverbs appear within a sentence
zero article	no article e.g. *Make love not war* (as opposed to *the war against terrorism*)

Index

A. Wallwork, *English for Academic Research: Grammar, Usage and Style*,
DOI 10.1007/978-1-4614-1593-0, © Springer Science+Business Media New York 2013

Made in the USA
Coppell, TX
18 March 2022

75209447R00151